A SOLUTION

TO **SOLUTIONS**

A PRACTICAL GUIDE TO UNDERSTANDING AND PREPARING SOLUTIONS IN BIOLOGICAL CHEMISTRY

FIRST EDITION

BY **T. MICHELLE JONES-WILSON, PhD**
EAST STROUDSBURG UNIVERSITY

cognella® | ACADEMIC PUBLISHING

Bassim Hamadeh, CEO and Publisher
Kassie Graves, Director of Acquisitions
Jamie Giganti, Senior Managing Editor
Miguel Macias, Senior Graphic Designer
John Remington, Senior Field Acquisitions Editor
Monika Dziamka, Project Editor
Brian Fahey, Licensing Specialist
Rachel Singer, Associate Editor

Cover image copyright © by Depositphotos / chepko.

Printed in the United States of America

ISBN: 978-1-63487-800-5 (pbk) / 978-1-63487-801-2 (br)

cognella® | ACADEMIC PUBLISHING

To my husband Paul and sons Ian and Isaiah for teaching me what is really important.

CONTENTS

ACKNOWLEDGMENTS

Thank you to my many students who inspired me via their errors to write this book. A special thank you to a fabulous student, Christopher Powers, for his invaluable editing and good humor.

INTRODUCTION
Why Are You Afraid of a Little Solution?

TO THE USER

The salutation above is to the user, not to the reader. That distinction is important because this book is highly practical. It is meant to be used; not just to be read. I have written this "text" (I hesitate to call it a text, as it is more of a handbook or guide) to provide a resource for students and technicians in the biological and biochemical sciences. Since I am an academic, I often refer to the *student*. However, *laboratory technician* can be as easily substituted. It is unlikely that there is any information in this book that you have not learned in one of your classes or laboratories. However, the old adage is true—if you don't use it, you lose it. It is frustrating to know that at one time you did a specific task and now can't remember how to do it. So, this book has been written to provide you with three things: first, a bit of background and theory related to different aspects of laboratory calculations and practices. Second, to provide a quick set of instructions, directions, and examples to help you accomplish selected particular laboratory calculations. Third, this book is intended to convince you that these laboratory tasks are simple and well within your abilities.

Each chapter and subchapter is divided into sections specific to a particular task or operation. Subchapters are further divided into two sections. The first section is a *background and theory* section that provides a brief description and reminder about the basic chemistry of each topic. Hopefully, these introductions to a topic will provide you with sufficient background to either help you remember what you previously learned or maybe allow you to understand the topic for the first time. The second section is *in the laboratory*, where practical examples and tips are provided specific to the task.

One thing I hope you will take away from this text is that the calculations and practices described herein are actually simple practices and calculations. Some may be multistep, but each step is simple. I am continually amazed at the minimal confidence students express in their own abilities to *do math* or *do chemistry*. There are no calculations in this book that are beyond the capability of an average high school student. But somewhere, you were probably told that math is hard and that this stuff is complicated—and thus you believed it. You have become afraid of a little solution!

Basic solution calculations are simple, but students yearn to complicate them. What I mean by this is that students insert extra terms and use extra equations because they are looking for a more complex or torturous route to the answer. Overthinking often competes effectively with not thinking at all. When I ask a student why he or she inserted a particular term and didn't just solve directly, the student often replies, "Well, I did it that way at first, but it seemed too easy." It seems as if students believe that if they can do something easily, it must mean it is wrong. I hope this book will convince you that nothing covered here is difficult if you are organized and follow the basic methods and logic outlined.

In addition, I have added a practice I call a *reality check* to calculation exercises. Too often, students or workers believe their calculators. They take the number that comes from the calculator window and use it without thought. If you make a mistake in an input or forget to divide by 1000 or multiply by 1000 as appropriate, you will obtain a numerical answer that simply doesn't make sense. If you calculate the volume of a cell as 1000 L, you might want to rethink your answer! This sounds funny, but I get answers like that all of the time. You must look at your number and determine whether it makes sense. In the 1000 L cell example, the student most likely multiplied by 10^6 somewhere where he or she should have divided by 10^6. If you look at your answer and think about it, you will see that you have made a mistake and can go back and check your work. Always check your answer against reality.

The style I have chosen in this work is the same one I use in the classroom: brief, direct, and with a dose of humor. Too often, my students approach laboratory calculations and preparation with dread. Possibly they feel intimidated? It seems that any time any mathematics is involved, a wall is erected. However, I hope this book will show the timid student or technician that the level of mathematics needed to accomplish any of these tasks is well within that individual's abilities. In addition, I have provided a chapter concerning basic algebra and logarithms, the two areas in which students struggle—probably because they simply don't remember it.

Also, I have written sections on using the Web and other resources as well as practical information regarding the correct equipment to use for a particular task and handling of biomolecules. All of the information in this book can be found elsewhere, either in your general chemistry text or in bits and pieces on the Web. However, there is no printed comprehensive resource that brings much of the basic information about laboratory practice and solution preparation together into one source so that it is easily accessible. Hopefully, this text will provide you with that resource.

CHAPTER 1
The Basics—How Do I Find the Data?

SECTION A

A Quick Primer to Using the *CRC*, the *Merck Index*, and What You Can Reliably Find on the Internet

In order to make many solutions, you will need to obtain physical data such as densities, formula masses, or other basic chemical information. The question is, where should you find it?

Students often rely exclusively on the Internet for information. In general, this approach is successful in finding basic chemical information. However, when you Google it, you must make sure that you choose specific and appropriate search terms to find the information you seek. There are also a number of free chemistry- or science-based search engines like ChemSpider (chemspider. com) and PubChem (http://pubchem.ncbi.nlm.nih.gov). Of course, if you are lucky you have access to paid databases like CAS (Chemical Abstract Services). But for basic information needed for laboratory calculations, the free databases will provide much if not all of what you need.

When using Internet resources, you MUST critically evaluate the information you obtain. Not all of it is created equal, and while much—some can argue, the majority—of the information obtained is correct, you can also find many errors. It is important to check the source of any data you obtain (and just as important to record the location). Look at the URL (universal resource locator, aka Web address) and judge its validity. In general, .edu sites and .gov sites such as vitalstatistics.gov can be relied upon for accurate data. Be more wary of .com sites; however, many are quite reliable. For example, an easy place to find basic chemical information is on the websites of chemical manufacturers like Sigma-Aldrich (http://www.sigmaaldrich.com). These companies list their catalogs and catalog descriptions that contain physical data such as molecular formulas, formula masses, densities, and melting and boiling points. Be careful that you are looking up the correct compound, as many chemicals, especially inorganics, can be sold as various hydrates or salts (if you have forgotten about hydrates and salts, see Chapter 3).

A good, reliable online source is the National Institutes of Standards and Technology Chemistry WebBook at http://webbook.nist.gov. You can search by common or IUPAC name, molecular formula or structure and obtain structures, formula masses, and basic physical data. Students often rely heavily on Wikipedia. Some faculty rage against Wiki, labeling it an inappropriate site. However, I disagree. Wikipedia is often a very good source to find basic information and to provide a user with background so that he or she can then look further in depth. The problem with Wiki, and the reason why some disregard it, is that it is a public site and is not screened, nor is the posted information verified. Information on Wikipedia is provided by users who register on the site. There is no guarantee that the person who wrote a particular topic is an expert or even knowledgeable in the field about which he or she wrote. However, most people who post are knowledgeable and the information provided is accurate. When using Wiki, you need to be wary. Never accept information gained from Wiki as correct without checking with another source. One way to evaluate information on Wiki is that much of it is referenced (at least, good information is referenced). Check out those references for validity. Wiki is a good first stop on the Web; however, it should never be the only and final destination.

In the age of the Web, printed sources have fallen into disuse in many ways. However, two books, the *CRC* and the *Merck Index*, are extremely valuable to have in the laboratory. Of the two, the *Merck Index* is probably more useful in the biochemical laboratory, and it is arguably a bit easier to use. Chemical information in the *Merck Index* is primarily geared toward pharmaceutical applications focusing on basic biological and organic molecules. The *Merck Index* is arranged alphabetically, with references for common names. It provides physical information about the compounds contained and a summary and references about preparation and uses. The *CRC*, like the *Merck Index*, is published every year. The *CRC Handbook of Chemistry and Physics* is by CRC Press. This tome contains physical information for hundreds of thousands of chemicals, as well as tables of physical constants and information of chemical processes, among other topics. The *CRC* is not as easy to use as the *Merck Index*, but it is more comprehensive. The *CRC* is divided into sections, and within those sections, information is generally listed alphabetically by structure. The trick with the *CRC* is determining what section to use. Also, when searching for information on chemicals, the *CRC* is now available in an online version (although the most up-to-date edition is not free). This version has the beautiful feature of a strong search engine. In addition, the latest versions allow the user to search by structure as well as by name.

SECTION B

Basic Terminology: Which One Is the Solvent?

In order to accomplish solution preparation, you need to have a basic command of the terminology associated with solutions. You can't find the correct data if you are using incorrect search terms! While the definitions may seem obvious, it is not always so and can be a point of confusion.

> *A solution is a mixture of two or more substances that combine in a single phase.*

> *A solution must be homogeneous throughout and can exist in any phase.*

Most often we experience liquid solutions. In the biochemistry and biology laboratories, most solutions are not only liquid solutions but are aqueous solutions, meaning water is used as the solvent. The solvent is the component of a solution that is present in the greatest quantity. Commonly, the solvent is a liquid. However, solvents can be gases in gaseous solutions. The solute (or solutes) is (are) the component(s) of a solution that is (or are) present in lower concentration compared to the solvent. Solutes can be of any phase—solid, liquid, or gas. A common example of a solid solution is an alloy like brass; a gaseous solution is all around us—namely, air.

In the life sciences, the most commonly encountered solution is a solid dissolved in liquid water. These solutions are noted with a subscripted (aq) to denote the phase. An aqueous solution is not a liquid, denoted with a subscripted (l). However, gases can be dissolved in water as well. The notation is still (aq) for an aqueous solution of a dissolved gas.

TRAP ALERT! Description of the dissolution of a solid as becoming a liquid is a common freshman mistake. The process of changing from solid to liquid is a phase change called melting. In solution formation, the solid dissolves; it does not melt. It is in fact still a solid, either ionic (disassociated into ions) or molecular (no ions); however, it has dissolved in the liquid to form a homogeneous mixture.

CHAPTER 2
Units

Units — You Shouldn't Live without Them

BACKGROUND AND THEORY

Measurements versus Numbers

Units are essential. You probably need convincing!

One problem with calculators is that they deal exclusively in numbers. Performing a calculation and obtaining the correct numerical answer is only part of the solution to any problem or task ahead of you in the laboratory. What is missing from the calculator are the units. Students are notorious for abandoning units in any calculation. Unfortunately, without the units, what you have is simply a number, which is most often utterly useless in the laboratory. What is critical for real work is a measurement.

For instance, what would you do if you were told to obtain 1.06 of substance A? You would ask, 1.06 what? Grams? Milliliters? Liters? In other words, you need the units! Without the units, you have no idea how to complete the assigned task.

5

A **measurement** consists of a number (the magnitude) and its unit. A measurement is useful in the laboratory.

The units are actually critically important in obtaining or correctly calculating the number. Rather than making units the bogeyman, you should make them among your best friends. If you walk hand in hand with units, you will be successful. So, cozy up! Somewhere in your educational past, you probably heard the term **dimensional analysis**. Don't quiver in fear: dimensional analysis is, in fact, quite easy, and when done correctly, it virtually guarantees a correct answer.

You may suffer from math phobia or math anxiety and are probably reliving with horror that general chemistry exam where calculations were piled upon calculations piled upon formulas. Well, put your fears to rest. If you can add, subtract, multiply, and divide (and yes, you can use your calculator for these tasks) and do some basic algebra, then you can use dimensional analysis. But since the term *dimensional analysis* may bring back nightmares, we can rebrand it as **unit cancellation**, followed by the even more fun **unit flipping**.

So, how much algebra is involved?

$$\frac{ab}{ac} = \frac{b}{c} \quad \text{the variable } a \text{ cancels out leaving } b \text{ and } c \quad \frac{\cancel{a}b}{\cancel{a}c} = \frac{b}{c}$$

If you can handle the above algebraic equation, you can conquer unit cancellation. If you forgot about the rules used in canceling variables (units) and other basic mathematical stuff, skip ahead to Chapter 4.

In the Laboratory

Putting unit cancellation and unit flipping to good use—complete persuasion!

At this point, you probably still don't believe me. You remember dimensional analysis as much more complicated than the simple equation above. Well, let me convince you that it is not. Let's say that

a represents mass in grams (g)

b represents moles (mol) and

c represents volume in liters (L).

Simply substitute the units for the variables shown above and perform the same task:

$$\frac{ab}{ac} = \frac{b}{c} \quad \text{substituting} \quad \frac{g * mol}{g * L} \quad \text{and} \quad \frac{\cancel{g} * mol}{\cancel{g} * L} = \frac{mol}{L} = M$$

The above example shows how you would obtain the concentration of a solution in mol/L or molarity (M) from the mass of a solute (g), the formula mass (often called molecular mass) of the solute (g/mole), and the volume of solution, L. Molarity is covered in gory detail in Chapter 5. So, skip ahead for a definition if you need one. But for now, let's think about this calculation and how you could accomplish it if you don't remember all of those formulas.

You are probably thinking, "So, what does all this algebra have to do with practical laboratory work?" Here's an example. To make a simple solution of a solid dissolved in a liquid (most commonly water in the biochemical laboratory), you generally weigh the solid obtaining its mass in grams. You now have variable *a*. If you know the molecular formula of the solid, you can obtain its formula mass by checking the Internet (see Chapter 1) or by adding the atomic masses from the periodic table in the correct ratios (skip ahead to Chapter 3 for a review of formula mass if you need to). Or, conveniently, most manufacturers print the formula mass right on the bottle! Mathematically, a formula mass is the ratio of the number of grams to one mole of that substance. More simply put, formula mass answers the question, "If I have a mole of substance A, how many grams does it weigh?" Its units are, therefore, g/mol or variables *a/b* in our example.

Moving back to the lab, once you have weighed the solid, you put it into an appropriate flask and fill to the mark with your solvent to obtain the final volume, in this example variable *c* in liters (L). To find the concentration of the solution you made in molarity (M), which has units of moles/L, you will need to employ the collected information to obtain your concentration.

You could memorize the general chemistry textbook recipe "to find the molarity of a solution, divide the mass by the formula mass and divide by the volume." That will, of course, work. However, you will have to memorize a formula for every different type of solution calculation and then make sure that you use the correct formula for the correct situation in order to obtain the right answer. That gets complicated, and it's what gives chemistry a bad reputation. Besides, your memory plays tricks on you. You might have survived general chemistry by memorizing from chapter to chapter. But you probably struggled on the comprehensive final exam. Even if you did well in general chemistry, if you don't use the formulas on a daily basis, you forget them, and probably worse, confuse them.

So how do you make up for your failing memory? You could use unit cancellation and unit flipping and never memorize a single formula! Sound tempting? It is all based upon the units. Here's how you do it.

Unit flipping—the key to success!

Let's examine the problem above in a systematic manner. If you need concentration in the most common unit, M, molarity (mole/L), you will need moles in the numerator (that's the top) and volume in the denominator (that's the bottom) of your unit. This comes from the definition of molarity (see Chapter 5).

Let's start with the numerator. You need moles. The collected available information, formula mass, contains moles in the denominator. However, you can "flip" that with no problem at all.

Any unit can be flipped. Mathematically, flipping is the same as dividing by or multiplying by 1 over the unit (that's its reciprocal in math-speak). Remember that when you flip the unit, you MUST keep the number with the unit. If you don't, that is where you will get into trouble.

Example 2-1:

Unit flipping is best illustrated with a practical example. Consider sodium chloride, NaCl, which has a formula mass of 58.5 g/mol. In other words, there are 58.5 g in one mole (58.5 g/mol), or one mole weighs 58.5 g (1 mol/58.5 g). As long as the number stays with its unit, a number and its units can be used right-side-up or upside-down.

This example can illustrate that units are essential because just flipping the number without flipping the units as well is incorrect. In other words, you know that 58.5 does not equal 1/58.5. If you only use the numbers without the units, you have changed the measurement.

In formula-speak:

$$\frac{58.5 \text{ g}}{1 \text{ mol}} = \frac{1 \text{ mol}}{58.5 \text{ g}} \neq \frac{58.5 \text{ mol}}{1 \text{ g}}$$

(*Note:* The 1 is added for clarity here; it is usually omitted)

Now you have moles in the numerator (mol/g) by using the formula mass "flipped." But you also need volume, L, in the denominator of molarity. That's easy. You have the volume of solution, and you can simply put it in the denominator by multiplying by 1/volume.

So far, our calculation looks like this

$$\frac{\text{mol}}{\text{g}} \times \frac{1}{\text{L}}$$

But in our final answer, molarity (M) has no mass (g) in it. No problem. In making your solution, you weighed the NaCl on the balance and recorded the mass in g. Inspecting the above calculation and using your basic algebra skills, you should see that if you multiply by the mass, g, will cancel, leaving only the units, mole/L.

$$\frac{\text{mol}}{\text{g}} \times \frac{1}{\text{L}} \times \frac{\text{g}}{1} \Rightarrow \frac{\text{mol}}{\cancel{\text{g}}} \times \frac{1}{\text{L}} \times \frac{\cancel{\text{g}}}{1} \Rightarrow \frac{\text{mol}}{\text{L}} \equiv M$$

So, unit flipping followed by unit cancellation (what you once called dimensional analysis in abject fear), tells you to take the mass of the solute and multiply it by 1/molar mass of the solute (aka divide by the molar mass), and then finish by dividing by the volume of the solution. This is the same action as prescribed by the textbook formula. However, using units required absolutely no memorization.

Example 2-1 with numbers:

What is the molarity of a salt solution made by dissolving 10.0 g of NaCl in 1.0 L of water? The formula mass of the solute NaCl is 58.5 g; 10.0 g are used, and the solution volume is 1.0 L.

$$\frac{1 \text{ mol}}{58.5 \text{ g}} \times \frac{1}{1.0 \text{ L}} \times \frac{10.0 \text{ g}}{1} \Rightarrow \frac{0.17 \text{ mol}}{\text{L}} \equiv 0.17 \text{ M}$$

Dimensional analysis is really quite simple, once you break it down. I hope that this example, and the dozens of others in subsequent chapters, will convince you that units are your friends; without them, you will be a very lonely, unsuccessful laboratory worker.

SECTION B

SI Units and the Inconsistent Others We Refuse to Abandon

BACKGROUND AND THEORY

What are SI units, and why do we need them?

So, just what are SI units? Around the time of the French Revolution, the Archives de la République in Paris began to standardize a system of measurements that are the basis of today's metric system. SI units, the common abbreviation for *le Système International d'unités*, later canonized as the International System of Units (SI) in 1960, are the result of those early and continuing efforts.

Despite America's continued reliance on many English measurements, the metric system has become the standard in the scientific community. This is for good reason; despite your struggles with the metric system in grade school, the metric system simply makes sense. The metric system uses a series of base units (seven in total, listed in Table 2-1) and a series of derived units (derived from the seven base units). The same prefixes, *milli, centi, kilo*, etc. (see Table 2-2 for definitions), are applied consistently throughout the system, whether volume, length, mass, or force is measured.

TABLE 2-1 SI Base Units[1]

Measurement	Unit	Abbreviation	English System Alternative
length	meter	m	in (inch), ft (foot), yd (yard)
mass	kilogram	kg	oz (dry ounce), lb (pound)
time	second	s	s (seconds, something is the same!)
electric current	ampere	A	
temperature	kelvin	K	°F (Fahrenheit)
amount of substance	mole	mol	
luminous intensity	candela	cd	

TABLE 2-2 SI Prefixes[1]

Prefixes Greater than 1		Prefixes Less than 1	
Name (symbol)	Factor (multiply by)	Name	Factor (multiply by)
yotta (Y)	10^{24}	deci (d)	10^{-1}
zetta (Z)	10^{21}	centi (c)	10^{-2}
exa (E)	10^{18}	milli (m)	10^{-3}
peta (P)	10^{15}	micro (μ)	10^{-6}
tera (T)	10^{12}	nano (n)	10^{-9}
goga (G)	10^{9}	pico (p)	10^{-12}
mega (M)	10^{6}	femto (f)	10^{-15}
kilo (k)	10^{3}	atto (a)	10^{-18}
hecto (h)	10^{2}	zepto (z)	10^{-21}
deka (da)	10^{1}	yocto (y)	10^{-24}

TABLE 2-3 Selected SI Derived Units Useful in the Chemical and Biological Laboratory[1]

Measurement	Unit	Common Alternative	Helpful Notes
volume	m^3	cm^3	$m^3 = L$ (liter); $cm^3 = mL$
wave number	m^{-1}	cm^{-1}	
density	kg/m^3	g/ml	magnitude of $g/mL = kg/m^3$ since both numerator and denominator are divided by 1000
specific volume	m^3/kg	ml/g	
concentration	mol/m^3	molarity (M)	definition of molarity

TABLE 2-3 (Continued)

SI Derived Units with Special Symbols		Base Unit Expression	Commonly Used Alternative
frequency	Hz (hertz)	s^{-1}	rpm (revolutions/minute) 1rpm = 16.67 Hz
force	N (newton)	$mkgs^{-2}$	dyn (dyne) 1dyn = 10^{-5}N
energy (heat)	J (joule)	Nm	c (calorie) 1c = 4.184 J BTU (British thermal unit) 1 BTU = 1055 J
electrical charge	C (coulomb)	sA	
electric potential	V (volt)	$m^2kgs^{-3}A^{-1}$	
electrical resistance	Ω (ohm)	$m^2kgs^{-3}A^{-2}$	
capacitance	F (farad)	C/V	
temperature	°C	K − 273.15	°F—see below for temperature conversions[2]
radionuclide activity	Bq (becquerel)	s^{-1}	Ci (curie) = 3.7×10^{10} Bq
absorbed dose	Gy (gray)	J/Kg	rad = 0.01 Gy
dose equivalent	Sv (sievert)	J/Kg	rem = 0.01Sv
catalytic activity	kat (katal)	s^{-1}mol	U (enzyme unit) ≡ mass of enzyme that catalyzes 1 μmol of substrate/min = 1/60 μkat

[1]A complete list and explanation of SI units and their historical derivations can be found on the Web at the National Institute of Standards and Technology (NIST) website (http://physics.nist.gov).

[2]A note about the Celsius and Kelvin degrees: The Celsius scale is more commonly used in the biochemical laboratory than the Kelvin scale, which is most commonly used in thermodynamic calculations. The Kelvin scale is defined by reference to the ice point, which equals 273.15 K, or 0°C (the degree mark is not used with the K unit). The Celsius degree is equal in magnitude to the Kelvin degree; the scale has "slid" by 273.15 units. Thus, when using changes in temperature, the magnitude or size of Δ°C is the same as ΔK. So, if you need a temperature difference, you do not need to convert Celsius to Kelvin to obtain the difference. This is not true, however, of Δ°F. For the Fahrenheit scale, the magnitude or size of a degree is different from the Celsius and Kelvin scales. Changes in temperature measured in the Fahrenheit scale (°F) cannot be directly applied to the Celsius or Kelvin scales. Instead, the Fahrenheit temperatures must be converted to their Celsius or Kelvin equivalents and then any difference in temperature calculated.

$$°F = \left(°C \times \frac{9}{5}\right) + 32$$

This makes much more sense than 12 inches in a foot followed by 3 feet in a yard, or even worse, 5,280 feet in a mile!

While the English system may be familiar to us and historically quaint, it isn't the easiest to manipulate in the laboratory. Studying the basis of the English system is an interesting history lesson. Who doesn't think that a system of length based upon three barleycorns laid end to end (the definition of an inch) or the gap between the king's nose and his outstretched hand (the definition of a yard) are interesting ways to define measurements? But to be fair, the English system has been improved and standardized over the years. For instance, an act of Parliament under Elizabeth I standardized the distance of a mile as 5,280 feet. However, it still has "miles" to go! The mile is a perfect example. In the United States and England, a mile is 5,280 feet. But in Austria, a mile is 8,297 yards, and in Italy, it is 2,025 yards, and so on from country to country. Plus, the English system relies on those pesky things called fractions. No simple movement of a decimal place for the English! No, instead divide by 2 or 3, or how about 16! If you find yourself resisting the metric system and SI units, just think about how much easier it is to determine the number of milliliters in a liter than the number of cups in a gallon!

In the laboratory, SI units are the norm for measurement of mass and volume and thus for concentrations. In addition, the SI system uses standardized prefixes that change in multiples of ten, making conversion easy by simply sliding the decimal point. That is, as long as you slide in the correct direction—see later in this section for guidance in that simple task!

In the Laboratory

How much is it? Getting a practical feel for the units of the metric system.

Because those of us living in the United States often use English units in our daily lives, we don't often think about the magnitude of an SI unit. So, if we are told to weigh 50 g, we don't necessarily have a good feel for just how much that is. Learning to relate the SI units to the units you use in your everyday life will help you avoid mistakes in the laboratory. For example, I tell my students that a first-class letter is mailed for one stamp if it weighs an ounce or less. We all know about how much a letter weighs; it is about the mass of a standard envelope and four or five sheets of paper. One ounce is 28.3 g. Thus, a gram is considerably smaller than an ounce, and your 50 g mass is nearly the mass of two letters. Most of us are more familiar with volume equivalencies due to the 2 L soda bottle. We know how much milk is in a gallon of milk and that a soda bottle contains a bit less soda. In fact, there are 2.4 liters in a gallon; about 17 percent less liquid is in that 2 L bottle of soda than in the gallon of milk.

Another area where we have problems is with temperature. We have no problem "thinking in Celsius" in the laboratory, knowing that 25°C is room temperature and 37°C is human physiological temperature. But if someone told you that the low would be 15°C today, could you decide whether you need to bring a sweater? A rough but fast mental conversion from °C to °F is to double the °C and add 32. So, our 15°C day is about 62°F (the exact value is 59°F)—and you just might need that sweater.

Sliding the decimal point correctly

One common problem in applying the metric system correctly in calculations is moving the decimal point in the correct direction. I have graded hundreds of papers where an answer is incorrect because a student converted a liter to a thousandth of a milliliter rather than a thousand milliliters. There are two approaches to conquering this problem. The first approach is mathematical. Use the unit cancellation method. Rather than trying to remember the text book formula "to convert from liters to milliliters, divide by 1000, or to convert from milliliters to liters multiply by 1000," use units and the definition of liters and milliliters.

For example, to convert 15 mL to L, try the following:

$$15 \text{ ml} \times \frac{1 \text{ L}}{1000 \text{ mL}} \Rightarrow 15 \text{ mL} \times \frac{1 \text{ L}}{1000 \text{ mL}} = 0.015 \text{ L}$$

In this case, the ratio 1 L/1000 mL is used so that the mL will cancel, leaving the desired unit L. Therefore, when the math follows the units, you divide by 1000. The reverse would be true if you wanted to convert from 0.015 L to mL. You would simply flip the conversion factor. One liter equals 1000 mL is the same as 1000 mL equals one L.

$$0.015 \text{ L} \times \frac{1000 \text{ mL}}{1 \text{ L}} \Rightarrow 0.015 \text{ L} \times \frac{1000 \text{ mL}}{1 \text{ L}} = 15 \text{ mL}$$

Using your units will ALWAYS work. If the units cancel correctly, then the numbers will follow.

The second approach is one of common sense. Just think about it! Which is smaller, a milliliter or a liter? If you are given liters and you need to find the number of mL, then you should get "more" mL (a greater number) than L. Dividing by 1000 will not give you more; it will give you *less*. Therefore, you must multiply by 1000. The number you get from the calculator is useless if you don't think about it and make sure that it makes sense. Always do a reality check with every calculation.

CHAPTER 3
Chemistry Basics

A Mole Isn't Just a Small Rodent

BACKGROUND AND THEORY

What is a mole, and why do I need to know about it?

Many of you remember the term **mole** from your general chemistry background. You may even remember Avogadro's number, 6.023×10^{23}. However, you may not remember or ever have known what a mole is and why it is important.

The mole defines the amount of a substance

First, a mole (mol) is the SI unit for amount of substance. A mole is a specified number of articles (in this case, 6.02×10^{23} is the number). Because atoms and molecules have different subatomic or atomic makeup, respectively, they must necessarily have different masses. For instance, a mole of hydrogen atoms, containing a simple proton and electron per atom, must weigh considerably less than a mole of carbon atoms, containing 6 protons,

6 neutrons, and 6 electrons in each atom. A practical, more visual example is to imagine a mole of lemons and a mole of grapefruit. You have the same number of each fruit. But since each grapefruit is significantly larger than each lemon, the mole of grapefruit would weigh considerably more. A mole relates to the number or numerical amount—not the mass—of a particular molecule or atom.

Why such a silly number?

You might wonder, why 6.02×10^{23}? This is an extremely large and inconvenient number. Avogadro's number (N_A) is so large it is difficult to imagine. One analogy is that the volume of all of the oceans on the earth is about 1.37×10^{24} mL, a little more than two times Avogadro's number. Another thing that really brings home the vast size of N_A is a rice analogy. In a good year, approximately 32×10^{12} kg of rice are produced on the entire planet Earth. If you assume a single grain of rice weighs about 25 mg (a reasonable assumption), it would take more than 470,000 years to produce Avogadro's number of rice grains. That's a lot of rice!

So, where does this number come from? Given the metric system's love for factors of ten, why not a simple thousand or million? Avogadro's number is based on chemistry. Avogadro's number is the number of carbon atoms in exactly 12 g of carbon-12 (the most abundant isotope of carbon). The concept of a mole was proposed by Amedo Avogadro (Avogadro's hypothesis) in about 1811. The number has been refined over years of study and is currently accepted as 6.022137×10^{23} and was named for Avogadro. The term mole reflects its molecular origin.

In the Laboratory

The mole is important in the laboratory because it standardizes formula mass (g/mol) (discussed in the next section) and is used in concentration determination (molarity, normality, and molality) discussed in Chapter 5. While you will rarely use Avogadro's number directly, understanding the concept of a mole and its magnitude is useful in making practical laboratory decisions.

SECTION B

How Many Names Are There for Formula Mass?

BACKGROUND AND THEORY

The molecular or formula mass (or molecular weight or formula weight) of a compound is determined by summing the atomic masses (or weights) of each element in a compound in the appropriate whole number ratios as defined by the molecular formula. Huh?

Essentially, formula mass (how we will refer to it in this text) is determined by adding up the atomic masses of each atom in a molecule. The atoms and number of each atom are determined by the formula of the compound.

In the Laboratory

Example 3-1:

As an example, the simple case of the molecule water, H_2O: the formula indicates that there are two atoms of hydrogen and one atom of oxygen in each molecule. Inspection of the periodic table (using the number found generally found below the atom symbol) provides the masses of each element. Strictly speaking, the mass is in amu (symbol u), or atomic mass units. However, for practical purposes, we read the mass as the mass in grams in a mole. This is possible because of the definition of a mole based on the mass of 12.0 g of C-12.

Therefore, the mass of hydrogen is 1.0 g/mol (rounded off) and of oxygen is 16.0 g/mol (also rounded off).

Thus, the formula mass of water is:

$$\left(2 \times \frac{1.0\ \text{g}}{\text{mol}}\right) + \left(1 \times \frac{16.0\ \text{g}}{\text{mol}}\right) = 18.0\frac{\text{g}}{\text{mol}}$$

It really is that simple.

Example 3-2:

As the formula becomes more complex, you simply add more terms corresponding to the number of atoms in the molecular formula. For example, for glucose, $C_6H_{12}O_6$ (the mass of C is 12.0 g/mol):

$$\left(6 \times \frac{12.0\ \text{g}}{\text{mol}}\right) + \left(12 \times \frac{1.0\ \text{g}}{\text{mol}}\right) + \left(6 \times \frac{16\ \text{g}}{\text{mol}}\right) = (72.0 + 12.0 + 96.0)\frac{\text{g}}{\text{mol}} = 180.0\frac{\text{g}}{\text{mol}}$$

This calculation can be done for simple molecules or molecules with hundreds or thousands of atoms.

A simple shortcut—look on the bottle!

Often, the formula mass for a compound is listed on the bottle or on the catalog description of the compound. However, some manufacturers provide a number only, without units. Make sure to add the units when you record the information.

Once again, a word about units!

Unfortunately, the periodic table just provides numbers, not units, and students sometimes forget that the correct units for formula mass are g/mol. This is essential to remember if you need to calculate concentrations or convert one concentration to another (Chapter 5). So, like always, units are critical here as well!

SECTION C

The Mystery of Hydrates and Salts Solved

BACKGROUND AND THEORY

What do you do about formulas that have "dots" in them? Generally, these compounds contain waters of hydration in their crystal lattice. These compounds are called hydrates, and the formula is written generally as $\cdot H_2O$; for example, $CuCl_2 \cdot 2H_2O$ (read cupric chloride dihydrate). A hydrate is a solid compound that contains a definite ratio of bound water within the structure of the molecule. Calculating the formula mass of a hydrate is simple. The water molecules are just added to the total mass. In the case above, the 2 in front of the H_2O indicates that two molecules of water are contained in each molecule of the compound.

Ionic solids are commonly referred to as salts. The salt with which you are most familiar is NaCl, or simple table salt, that you use to flavor your food and potentially raise your blood pressure. But in more general terms, in chemistry, a salt can mean simply an ionic solid. The term salt is also used routinely in buffer solutions to refer to the ionic solid that is the conjugate base of a weak acid. For example, a sodium acetate buffer $(NaOCH_3)$ is composed of a mixture of the weak acid acetic acid $(HOCH_3)$ and its salt $(NaOCH_3)$. In solution, the salt will exist as an ionic solute, disassociating into the ions Na^+ and OCH_3^-. In this case, the formula mass of the salt is higher than the acid because a hydrogen atom is replaced by a sodium ion.

In the Laboratory

Example 3-3:

Calculate the mass of our copper (II) chloride dehydrate salt.

The formula is $CuCl_2 \cdot 2H_2O$. The formula mass of Cu is 63.5 g/mol, Cl 35.5 g/mol, H 1.0 g/mol, and O 16.0 g/mol. Two water molecules contain 4 hydrogens and 2 oxygens (the multiplier 2 is applied to both the hydrogen and the oxygen). The formula mass of the compound is therefore:

$$\left(1 \times \frac{63.5 \text{ g}}{\text{mol}}\right) + \left(2 \times \frac{35.5 \text{ g}}{\text{mol}}\right) + 2 \times \left[\left(2 \times \frac{1 \text{ g}}{\text{mol}}\right) + \left(1 \times \frac{16 \text{ g}}{\text{mol}}\right)\right] = 169.5\frac{\text{g}}{\text{mol}}$$

You can simplify things by remembering that water has a mass of 18 g/mol (through use, you will quickly cement that fact in your mind); therefore, the calculation simplifies to:

$$\frac{1 \times 63.5 \text{ g}}{\text{mol}} + \frac{2 \times 35.5 \text{ g}}{\text{mol}} + 2 \times \frac{18 \text{ g}}{\text{mol}} = 169.5\frac{\text{g}}{\text{mol}}$$

Generally, whether a hydrate of a molecule or an anhydrous (without water) compound is used is not relevant to the chemistry of the compound. This is especially true when the substance will be dissolved in water to make a solution. The water within the crystal structure of the solid may be released and will simply occupy negligible volume with the solvent. What is important is to make sure that you are using the correct formula mass for the correct hydrate or anhydrous solid and that when considering the salt you include the mass of the cation.

SECTION D

Just What Is a Dalton, Anyway?

BACKGROUND AND THEORY

A dalton is a unit of mass equal to 1/12 the mass of carbon 12, which is assigned a mass of 12. This is the modern definition. Translated to English, that means a dalton is equivalent to an atomic mass unit. So, hydrogen has an amu (atomic mass unit) of one. In larger and more practical molar terms, hydrogen has a mass of 1 g/mol, or 1 dalton. The dalton is not an SI unit; it is most commonly observed when biological samples (especially proteins and nucleic acids) are part of the task. Often, the mass of proteins is expressed in daltons, as in the mass of hemoglobin is approximately 68,000 daltons. That is the same as stating 68,000 g/mol.

In the Laboratory

Formula masses are important in the laboratory because they are essential in calculating the number of moles from a mass or determining the mass to obtain when making a volume of specified concentration. Remembering the correct units (and using the correct units) is important in successful calculations and thus successful experiments. Daltons should be converted to or expressed as g/mol for calculations.

CHAPTER 4
Mathematics, That Dreaded Discipline

SECTION A

Basic Algebra: Canceling Units

BACKGROUND AND THEORY

What you need to know about algebra to be successful in the chemical or biological laboratory can be summed pretty quickly. Just a few reminders, and your rusty "math brain" will once again function. The best place to start is with a few definitions and descriptions of algebraic methods.

A variable is a symbol that represents a number. In math class, you usually used letters such as n, t, or x for variables. In solution calculation tasks, we use units (mL, mol, g, etc.) as variables (Chapter 2 A). Variables are useful representations because mathematical operations can be performed on variables as if they were numbers. In determining the concentration of a solution or in dozens of other laboratory calculations, using the units to "lead the calculation" will result in success. Most importantly, variables can be reduced or simplified by division.

For example, using traditional variables:

$$\frac{2x}{x} \Rightarrow \frac{2\cancel{x}}{\cancel{x}} = 2 \quad \text{OR} \quad \frac{x^2}{x} = \frac{x \bullet x}{x} \Rightarrow \frac{x \bullet \cancel{x}}{\cancel{x}} = x$$

Using units:

$$\frac{g}{mol} \times mol \Rightarrow \frac{g}{\cancel{mol}} \times \cancel{mol} = g$$

It's the same thing—just more letters!

Understanding exponents is also important in order to perform laboratory calculations. Exponents are used in scientific notation.

For instance, $1.0 \times 10^2 = 100$ or $1.0 \times 10^{-2} = 0.01$.

When used with variables, exponents can denote a simple multiplication, where:

$$x^2 = x \times x$$

as shown above. It is important to remember that when a negative exponent is used with a variable or units, the expression is equal to the reciprocal;
for example,

$$x^{-2} = \frac{1}{x^2}$$

or

$$mol^{-1} = \frac{1}{mol}$$

SECTION B

The Logarithm: A Review

BACKGROUND AND THEORY

Other than some simple algebra and basic math, the other mathematical trick you will need up your sleeve is the ability to use logarithms. Mostly, you will simply use your calculator and "plug and chug" and obtain the value of a log. However, remembering the basic rules of logarithms will keep you from making common mistakes and may help simplify some calculations.

Base 10 logarithms

Just what is a logarithm? The logarithm of a number is the power, or exponent, to which the base must be raised in order to produce a number. In plain English, the common logarithm—the base 10 logarithm—is the one most often used in the laboratory (base e is discussed later). This is especially true in the life sciences, where pH, a common logarithm, is essential.

The common logarithm of a number is simply the number of times 10 must be raised (or multiplied by itself) to obtain the original number. This is understood most easily by studying a numerical example:

$$\log_{10}(100) = 2$$

because:

$$\log_{10}(100) = \log 100 = \log 10 \times 10 = \log 10^2 = 2$$

Two is the exponent, or the number of times 10 is multiplied by itself, to obtain the value 100. Therefore, 2 is the solution to the logarithm.

And likewise:

$$\log 1000 = \log 10 \times 10 \times 10 = \log 10^3 = 3$$

Three is the exponent, or the number of times 10 is multiplied by itself, to obtain the value 1000. Therefore, 3 is the solution to the logarithm.

And for numbers less than 1:

$$\log 0.0001 = \log 0.1 \times 0.1 \times 0.1 \times 0.1 = \log 10^{-4} = -4$$

Note: The mathematicians will note that the base being used should be subscripted to the right of the log. The common logarithm should be expressed as $\log_{10}(100)$, for example. However, when base 10 is used, the subscript is routinely dropped, and log100 is written instead.

The logarithms of simple whole numbers of the base (in our case, 10) are relatively easy; you might realize from the above examples that the number of zeros equals the value of the log. This is true for all integer multiples of the base 10. So, these logarithms can be accomplished easily in your head. No calculator needed!

What about logarithms of numbers that are not simple multiples of the base? People used to look up the values in log tables. However, the calculator (or spreadsheet) comes in handy for this task today.

   For example:

$$\log 1052 = 3.02$$

and

$$\log 9009 = 3.95$$

I got the above numbers by plugging the values into my trusty calculator. However, by understanding the logarithm, I can make sure my answers make sense—a reality check. This can avoid a simple mistake that can be carried throughout a string of calculations.

Reality Check:
1052 is fairly close to (but a bit greater than) 1000. The log of 1000 is 3 (three zeros). Consequently, our logarithm of 1052 should be slightly more than 3. And lo and behold, it is 3.02.

Also:
9009 is close to 10000. The log of 10000 is 4. So, the logarithm of 9009 should be close to (but less than) 4; it is 3.95.

   To be successful in the laboratory, you should think about the numbers you are using and make sure they make sense in the context of your task. Always perform a reality check before making a measurement.

The antilog is not a group of people who oppose logarithms

The antilog is the inverse function of a logarithm.
   For instance, since the log of 100 is 2, the antilog of 2 is 100. The antilog is expressed exponentially as 10^x. If $10^x = 100$, then $x = 2$. The antilog is 10 raised to the second power, which is 100.
   The antilog function is usually the second function of the log button on your calculator. It comes in handy for less straightforward situations like the antilog of 3.03, which is $10^{3.03} = 1078$. That calculation is not as easily done in your head.

Reality Check:
$10^3 = 1000$ and 3.03 is slightly more than 3. Therefore the answer to $10^{3.03}$ should be more than 1000, and 1078 is more than 1000.

It's all natural, the base e

Natural logs, written ln, are not base 10 logarithms: they are base *e* logarithms, and base *e* is approximately 2.7. A natural log, ln *x* is really $\log_{2.7}(x)$. And the inverse function is e^x.

What do you really need to know about ln? Since we don't function day to day in a base *e* world, calculations with this base are not intuitive, and the tricks above shown for base 10 aren't as relevant. The most important thing is that it is not the standard base 10 log, so don't assume it is and use the log key on your calculator. Otherwise, ln comes up in a few places in laboratory calculations; simply make sure you use the correct function on your calculator or spreadsheet.

The log rules

Mathematical Operations and Logarithms

You can add, subtract, multiply, and divide logarithms. You just need to remember the rules, and luckily, they are simple. If two numbers or variables are products (multiplied together), then the logarithm of the product is equal to the sum of the individual logarithms. This is more understandable in equation form:

$$\log(a) + \log(b) = \log(a \cdot b)$$

For division/subtraction:

$$\log(a) - \log(b) = \log(a/b)$$

This makes realistic sense in our base 10 world. You can remember the expressions by thinking about a simple numerical example.
If both *a* and *b* are 10, then:

$$\log(a) + \log(b) = \log(a \cdot b)$$
$$\log(10) + \log(10) = \log(10 \cdot 10)$$

and we know from the discussion of logarithms that log 10 = 1 and log 100 = 2:

$$1 + 1 = \log 100 = 2$$

We know from our earliest days that $1 + 1 = 2$.

Additionally, it is often handy to remember the rules of exponents as applied to logarithms. If the log of a number or variable is raised to an exponent, then the numerical answer is equivalent to the exponent times the value of the logarithm of the variable or number. In equation-speak:

$$\log (a^b) = b \cdot \log (a)$$

to put this in line with our examples above

$$\log (10^2) = 2 \cdot \log (10) = 2$$

We know this is correct because 10^2 is 100, and the logarithm of 100 is 2. This simple example should help you remember the mathematical relationship.

Where does the negative go? The problem with pH

This may seem like a strange question, but it is one that comes up often in pH calculations. When calculating pH (see Chapter 7), you take the common log of the molar hydronium ion concentration and multiply it by negative 1.

$$pH = - \log [H_3O^+]$$

When you need the hydronium ion concentration from the pH, you need to remember your exponent rules. The −1 in front of the pH expression comes into the log expression following the rules for manipulating logarithms. The −1 is a multiplier, so it becomes an exponent. Thus, to obtain the molar concentration from the pH, you must take the antilog of −1 times the pH:

$$[H_3O^+] = 10^{-pH}$$

There are many other bases used in different applications; for example, you are probably familiar with base 2, or binary code, used in computing. There are also rules for the treatment of imaginary and complex numbers. But we don't generally need these in the biochemical or biological laboratory, and they aren't covered in this handbook.

In the Laboratory

Where will you most likely encounter logarithms? Of the base 10 variety, you will need to master logs to calculate pH or determine hydronium ion concentration (H_3O^+) from pH (Chapter 8 C). You will also need to be able to manipulate logarithms to complete buffer calculations (Chapter 8 D).

CHAPTER 5
Concentration Units

SECTION A

Common (and Not So Common) Concentration Units: When, How, and Why?

BACKGROUND AND THEORY

The main focus of this work is to help students in the biological and biochemical laboratories be confident in making solutions. Consequently, this chapter begins to attack the most common—and a few not so common—solution concentration units you will encounter. It also provides examples of how you go about making one solution from another solution when the concentrations of the two solutions are listed in different units. This chapter also provides the background you will need to master making buffers (Chapter 8), performing dilutions (Chapter 7), and the seemingly daunting, but really simple, task of setting up a multicomponent assay (Chapter 9).

The most commonly used unit in the laboratory to express concentration is the SI unit of molarity (moles/L, abbreviated M) (see Chapter 2 for SI base and derived units). However, you will often encounter non-SI expressions of concentration like molality (*m*), normality (N), parts per million (ppm), or sometimes percentage solution. The use of and potential problems with percent concentrations are discussed in part C of this chapter.

However, any of the above listed units are preferable to the worst-case scenario: no listed concentration and units. It is critical that all solutions (and all samples, for that matter) be appropriately labeled with the contents, concentration, date, and often the user/maker's name or initials. Good labeling practices will not only serve you well with notebook audits but also will save you time. If you keep good records and label solutions so that you can find the record, you need do a calculation only once. When you need to remake a solution, you can simply follow your previous protocol— as long as you can find it!

SECTION B

Activities—the Right Way to Do Them (Generally, We Ignore Them)

The activity and/or activity coefficient must be determined for the particular solute in the particular solution under the specified conditions. Activity is important for solutions, especially aqueous solutions containing ions, because ions interact with each other in solution, effectively changing the behavior of the solution. In essence, when we measure pH, we are not really measuring the concentration of hydronium ion; we are actually measuring the activity of hydronium ions. In fact, every calculation done using concentration rather than activity is strictly incorrect. When we make buffers and ignore the ionic strength of the solution by using concentration terms and the Henderson-Hasselbalch equation (Chapter 8 F) rather than activities, we will commonly find that the predicted pH differs from the actual pH measurably. Thus, even a well-done buffer calculation followed by a carefully measured preparation will suffer because concentrations rather than activities were used. Buffer pH will need to be adjusted to hit the mark intended.

So what is the difference between the concentration we generally use for solutions and the activity of the solution? And why don't we just use activity since activity is really the most accurate and correct term?

It is easier to answer the second question first. Activity and activity coefficients can be devilishly hard to measure, and for the most part, in dilute solutions of low ionic strength, the activity coefficient is very close to one and the activity is almost equal to the concentration. Therefore, the error imparted in most common laboratory practices, outside of careful analytical measurements, by using molar concentration instead of activity is generally small and is neglected. However, understanding that concentration differs from activity is important and explains much of the variations seen in aqueous solutions between calculated and observed behavior.

The first question posed addresses the difference between concentration and activity, or effective concentration. What comes next is a lot of theory. When we use concentration terms, we are stating that the concentration of the solute is equal to the activity of the solute in the solution. When we list the concentration of a component of a solution, we are making the implicit assumption that the solution is an ideal solution. You probably remember what an ideal gas is, but might not know about ideal solutions. There is little practical difference other than the phase.

The ideal world

In an ideal gas, we assume that each particle (atom or molecule) is independent of every other particle in the gas. In other words, molecules aren't friendly; there are no intermolecular (molecule to molecule) interactions. The particles in an ideal gas do not repel or attract each other in any way; each is independent of all others. In an ideal gas, the activity of a component is one (from Henry's law). In an ideal solution, we make the same assumption. We assume that each molecule of solute is independent of every other molecule of solute. In an ideal solution, we have no aggregation of solute molecules and no interaction between the molecules. Now, if you are thinking, you will know that this is not really true in solutions. If you have an electrolytic (charged) solute, it will repel and attract other molecules, depending on charge. Solutions are generally not ideal, and when we make the assumption that each molecule or mole (group of molecules) acts independently, we impart error. Additionally, activities are temperature dependent. This should make sense, as the behavior of molecules (molecular movement) will vary with temperature. Of course, this throws yet another wrench into the works when using activities in the laboratory. The activity coefficient is a unitless factor that contains all of the non-ideality of the solution. So, if we really want to be very accurate, we should be using an activity coefficient (γ, as a multiplier of our concentration, to obtain the "true concentration," or activity, at a specified temperature.

But as stated above, activity coefficients can be difficult to measure and calculate accurately. The question arises, do we need to be that accurate? Luckily, the answer is very often no. Our savior is that most of the solutions we deal with are fairly dilute. In a dilute solution, interaction between solute molecules is minimized because they simply don't "run into each other" that often. This is why we say that solution ideality is realized at infinite dilution. If solutions are extremely dilute (infinitely so), then the molecules will not interact because they simply don't find each other. It's like playing bumper cars at the fair with only one car in the ring. For the most part, using non–activity corrected concentration terms to describe solution concentrations is just fine, and it is what you will generally find outside of the analytical or physical chemistry laboratories.

To be complete, and in case you ever need it, the mathematical definition of activity in terms of chemical potential (partial molar-free energy) and using the activity coefficient are provided below:

$$a_i = e^{(\mu i - \mu_i^0 / RT)}$$

where a_i refers to the activity of component i, μ is the chemical potential of component i at temperature T and μ_i^0 is the standard state chemical potential, R is the gas constant and T is the temperature in kelvin.

OR

$$a_i = \gamma_{c,i} \cdot c_i / c_i^0$$

where a_i refers to the activity of component i, $\gamma_{c,i}$ is the activity coefficient of component i, c_i is the molar concentration, and c_i^0 is the molar concentration under standard state conditions.

Given these equations, you will understand why we usually ignore activities in favor of the more easily measured concentrations.

SECTION C

Molarity (M)—the Most Common Solution Concentration Unit

BACKGROUND AND THEORY

The most common, and SI, unit used to measure concentration in the laboratory is molarity. Molarity is defined as the moles of solute per liter of solution (mol/L) and is abbreviated with a capital M.

In the Laboratory

Note that, for molarity, the volume in the denominator is liters of solution, not liters of solute. This is an important point when it comes to actually making the solution. The correct way to make a molar solution is to add the solute to the appropriate volumetric container (see Chapter 10), then add the solvent (usually water) to "the mark" to obtain the final correct solution volume. You should not measure the solvent volume separately and add to the solute. The error introduced for addition of the solvent to a small mass of solid will be small. However, the error introduced if the mass is large or if the solute is a liquid may be significant. Additionally, there is simply no need to introduce this error if you make the solution correctly.

Making an M Solution

If the solute is a solid, it is generally weighed and then the solvent added as described in the previous paragraph. To obtain the molar concentration, you use dimensional analysis to guide your calculation as illustrated in Chapter 2.

$$\cancel{g} \times \frac{mol}{\cancel{g}} \times \frac{1}{L} = \frac{mol}{L} = M$$

To calculate molarity from a measured solute mass and solution volume, the units lead you to divide the mass by the solute's formula mass (multiply by the flipped formula mass) and divide by the volume.

Example 5-1:

Calculate the molarity of a solution where 18.2 g of sodium chloride (formula mass 58.5 g/mol) are diluted to 500 mL with water.

$$18.2\,\cancel{g} \times \frac{1\,mol}{58.5\,\cancel{g}} \times \frac{1}{0.5\,L} = \frac{0.62\,mol}{L} = 0.62\,M$$

To calculate molarity obtained when the solute is a liquid, two approaches can be taken. The first is identical to that above for a solid. This is used if the mass of the liquid solute is obtained (yes, you can weigha liquid!).

Example 5-2:

Calculate the molarity of a solution composed of 10 g of absolute ethanol dissolved in water to make 250 mL of solution:

To solve this problem, the formula mass of ethanol is needed. This value is 46.0 g/mol.

$$10\,\cancel{g} \times \frac{1\,mol}{46.0\,\cancel{g}} \times \frac{1}{0.25\,L} = \frac{0.87\,mol}{L} = 0.87\,M$$

The second approach is used when the solute is measured volumetrically. In this case, you will need the density of the solute to obtain its mass. Remember, the density of liquids varies with temperature; therefore, if accuracy is required and the solute is measured volumetrically, temperature must be obtained and the density at that temperature used. The calculation has one more step, but by following the units, you should have no problem obtaining the correct solution concentration.

$$\cancel{mL} \times \frac{g}{\cancel{mL}} \times \frac{mol}{g} \times \frac{1}{L} = \frac{mol}{L} = M$$

Example 5-3:

For Example 5-3, recalculate the molarity if 10.0 mL of ethanol had been used.

To solve this problem, the density of ethanol is 0.789 g/cm^3 at room temperature (remember, a cm^3 is a mL; Table 2-3) is needed since the solute is measured by volume.

$$10.0 \, \cancel{mL} \times \frac{0.789 \, \cancel{g}}{\cancel{mL}} \times \frac{1 \, mol}{46.0 \, \cancel{g}} \times \frac{1}{0.25 \, L} = \frac{0.69 \, mol}{L} = 0.69 \, M$$

Examples 5-2 and 5-3 are nearly identical. The additional term needed to obtain the mass of the solute from the volume (density) is included in 5-3. Despite the additional term, using units will guide you to the correct answer.

Determining Mass or Volume of a Solute from a Known Molarity

Essentially, determining the mass of solute needed from a known solution molarity is simply the reverse of the above calculations.

For example, to obtain mass of solute starting with molarity:

$$\frac{\cancel{mol}}{\cancel{L}} \times \cancel{L} \times \frac{g}{\cancel{mol}} = g$$

In this example, dimensional analysis can be used to eliminate units not desired in the answer. In this process, the desired unit, g, is brought into the calculation by the formula mass.

To obtain volume of solute from molarity:

$$\frac{\cancel{mol}}{\cancel{L}} \times \cancel{L} \times \frac{\cancel{g}}{\cancel{mol}} \times \frac{ml}{\cancel{g}} = ml$$

In this example, the density of the solute is also required since it is a liquid and it is not being weighed. In this case, it is flipped from the traditional g/ml expression of density. It is important to note the difference between the first term and the third term of the equation above. The volume in the denominator of the molarity is the liters of solution. The density in the third term is the volume of the solute not of the solution.

Example 5-4:

What is the mass of the solute ethanol in 250 mL, a 0.87 M solution?

$$\frac{0.87 \text{ mol}}{\text{L}} \times 0.25 \text{ L} \times \frac{46.0 \text{ g}}{\text{mol}} = 10.0 \text{ g}$$

Example 5-5:

What is the volume of the solute ethanol in the same solution as example 5-4?

$$\frac{0.87 \text{ mol}}{\text{L}} \times 0.25 \text{ L} \times \frac{46.0 \text{ g}}{\text{mol}} \times \frac{1 \text{ ml}}{0.789 \text{ g}} = 10.0 \text{ ml}$$

TRAP ALERT! Occasionally, you will be required to make a solution that is a certain molarity in a particular ion; for example, sodium ion or acetate ion. If this is the case, you must use the molecular formula to determine the number of moles of the ion per mole of the compound. (For instance, if you were asked to make a liter of an aqueous solution that is 0.5 M in acetate ion.) If you are making a solution with a particular concentration of the ion, it would matter whether you were using sodium acetate (NaOAc) or calcium acetate (CaOAc$_2$). Acetate ion has the formula CH$_3$COO$^-$ and is often abbreviated $^-$OAc.

Example 5-6:

Beginning with 0.5 M CaOAc$_2$ solution, calculate the mass of acetate ions in 1 L of solution.

For calcium acetate:

$$\frac{0.5 \text{ mol OAc}}{\text{L}} \times 1 \text{ L} \times \frac{158 \text{ g}}{\text{mol Ca(OAc)}_2} \times \frac{1 \text{ mol Ca(OAc)}_2}{2 \text{ mol OAc}} = 39.5 \text{ g Ca(OAc)}_2$$

Example 5-7:

Repeat Example 5-6 using NaOAc as the acetate ion source.

$$\frac{0.5 \; \cancel{\text{mol OAc}}}{\cancel{\text{L}}} \times 1 \, \cancel{\text{L}} \times \frac{82.0 \; \text{g}}{\cancel{\text{mol NaOAc}}} \times \frac{1 \; \cancel{\text{mol NaOAc}}}{1 \; \cancel{\text{mol OAc}}} = 41.0 \; \text{g NaOAc}$$

In Example 5-6, there two moles of acetate ion for every one mole of calcium acetate. Therefore, each mole of $CaAc_2$ provides twice as many acetate anions as each mole of NaOAc provides. In cases where you are asked for molarity of ions, it is helpful to use very specific units. In the example, above, rather than simply using mol/L for molarity, the unit was made more specific, mol $^-$OAc/L, as a reminder that the molar ratios must be considered.

Really, really small concentrations

The Millimolar and Micromolar Concentrations

Although not strictly an SI unit, you will often find concentrations listed as millimolar (mM) or micromolar (µM), particularly when working with biological samples.

An mM concentration is one millimol of solute per liter of solution, or 10^{-3} mol/L. Molarity, M, is one thousand times more concentrated than mM.

$$M \Rightarrow \frac{\text{mol}}{\text{L}} \times \frac{1000 \; \text{mmol}}{\text{mol}} = \frac{1 \; \text{mmol}}{\text{L}} = \text{mM}$$

Similarly, a micromol is one micromol of solute per liter of solution, or 10^{-6} mol/liter. Molarity, M, is one million times more concentrated than µM.

$$M \Rightarrow \frac{\text{mol}}{\text{L}} \times \frac{10^6 \; \mu\text{mol}}{\text{mol}} = \frac{1 \; \mu\text{mol}}{\text{L}} = \mu\text{M}$$

One common mistake is to express a millimolar concentration as millimoles per milliliter of solution. This is not millimolar; it is molar. In this case, you have simply multiplied both the numerator and denominator by 1000. Thus, you have multiplied by 1.

$$M = \frac{\text{mol}}{\text{L}} \times \frac{1000 \; \text{mmol}}{\text{mol}} \times \frac{1 \; \text{L}}{1000 \; \text{mL}} = \frac{\text{mmol}}{\text{mL}}$$

Molar concentrations can be expressed as mol/L or mmol/mL or µmol/µL. This alternative expression of molarity can be used as a convenient shortcut in a calculation.

SECTION D

Normality (N)—N Isn't a Typo

BACKGROUND AND THEORY

Occasionally, you will see a solution concentration expressed as **normality** using the unit abbreviation N. Students often think the N is simply an M gone wrong since the letters are adjacent on the keyboard. And many times the value of N is equal to the value of M, so you never realize your error. However, when the two aren't equal, it can cause significant problems. You need to know the difference between N and M.

Normality is defined as the ratio of the mass of the solute in grams to its gram equivalent mass per liter of solution:

$$N = \frac{g \text{ (solute)}}{\left(\text{equivalent}\right) g} \times \frac{1}{L}$$

The definition of equivalency depends on the type of chemical reaction being considered. This sounds more complicated than molarity because it *is* more complicated. That is one reason why normality is not used as often any more. In fact, the NIST (the SI guys, Chapter 2) has stipulated that the unit is obsolete and should no longer be used. Thus, it is often not taught in general chemistry courses. The big problem with that, however, is that it is still being used in many areas, and if you don't know about it, you can make some big mistakes.

In the Laboratory

The most common use for normality today is in Brønsted-Lowry acids and bases. This is the category of equivalency with which you must be most concerned. In fact, many companies still sell acids and bases with N, rather than M, concentrations listed. The use of normality is still common in clinical settings.

For an acid, one equivalent of acid is equal to one mole of protons; for a base, one equivalent of base is equal to one mole of hydroxide ions. Thus, the more confusing definition of N above can be refined:

$$N = \text{equivalents} \times \frac{1}{L}$$

Normality is actually quite easy to use, and I suggest that you simply add it as a term to your molar calculations.

$$\frac{eq}{mol} \times \frac{mol}{L} = N$$

This is easy to do in practice. For instance, if you have a monoprotic acid like HCl, you have one mole of hydrogen ions, or one equivalent of acid, per mole of acid. Your normality relates to your molarity, M, as follows:

$$\frac{1 \text{ eq}}{1 \text{ mol}} \times \frac{\text{mol}}{\text{L}} = N$$

In the case of a monoprotic acid (or single hydroxide-containing base like NaOH or KOH), the normality is equal to the molarity (this is how students have often confused N and M and luckily gotten away with it!).

However, for a diprotic acid like H_2SO_4, you need to take into account that two moles of hydrogen ion (two equivalents of acid) are available for each mole of acid.

$$\frac{2 \text{ eq}}{1 \text{ mol}} \times \frac{\text{mol}}{\text{L}} = N$$

Therefore, the value you obtain for normality will be twice the value for molarity. Normality is quite easy to use, but you need to be aware of the chemistry to avoid potentially costly mistakes in the laboratory.

Example 5-8:

The protocol calls for 500 mL of a 0.1 N solution of NaOH. Describe how the solution should be made.

To make this solution, you need the formula mass of NaOH (40 g/mol) and the ability to recognize that there is one equivalent of hydroxide per mole of NaOH.

$$\frac{0.1 \; \cancel{\text{equiv}}}{\cancel{\text{L}}} \times \frac{1 \; \cancel{\text{mol}}}{1 \; \cancel{\text{equiv}}} \times \frac{40.0 \text{ g}}{\cancel{\text{mol}}} \times 0.5 \; \cancel{\text{L}} = 2.0 \text{ g}$$

To make this solution, weigh 2.0 g NaOH into a container and dilute to 500 mL.

Example 5-9:

If the protocol from Example 5-8 called instead for 500 mL of a 0.1 N solution of $Mg(OH)_2$, describe how the solution should be made.

You will still need the formula mass of the solute (58.3 g/mol) and the ability to recognize that there are two equivalents of hydroxide per mole of $Mg(OH)_2$.

$$\frac{0.1 \ \cancel{equiv}}{\cancel{L}} \times \frac{1 \ \cancel{mol}}{2 \ \cancel{equiv}} \times \frac{58.3 \ g}{\cancel{mol}} \times 0.5 \ \cancel{L} = 1.46 \ g$$

To make this solution, weigh 1.46 g $Mg(OH)_2$ into a container and dilute to 500 mL.

SECTION E

Molality (*m*) — the Little *m*

BACKGROUND AND THEORY

Molality is another term used for concentration. Molality is defined as the moles of solute/kg of solvent and has the abbreviation *m*, usually italicized.

While not an SI unit and not as common as molarity (M), molality is used in the laboratory. One advantage of molality (*m*) is that it is not a temperature-dependent quantity. Since molarity (M) depends on volume (the denominator is the volume of solution) and volume changes with temperature, the value of M changes with temperature. Since you are generally working at or near room temperature in the laboratory, the changes in solution volume are slight and are usually insignificant and overlooked. Occasionally, however, temperature effects are important, and molality is the appropriate term to use. Sometimes, you don't really care about temperature effects, but your protocol is listed in *m* terms. In that case, you need to know how to use the *m* term and often to convert it to M terms (this chapter, Part I).

In the Laboratory

In practical terms, the difference between making an M and its little cousin *m* solution is simple. When making a molar, M, solution, you add the solute to the flask and dilute to the mark, effectively measuring the total solution volume (solute and solvent together). When making a molal, *m*, solution, you weigh the solute and solvent separately and then combine the two components. There is no measurement of the entire solution mass or volume.

Example 5-10:

A protocol calls for making artificial seawater. You need about 1 L, and the concentration is listed as 0.51 m NaCl.

For this problem, you will need the formula mass of NaCl (58.5 g/mol).

$$\frac{0.51 \; \cancel{\text{mol}} \; \text{NaCl}}{\text{kg water}} \times \frac{58.5 \; \text{g NaCl}}{\cancel{\text{mol}} \; \text{NaCl}} = \frac{29.8 \; \text{g NaCl}}{\text{kg water}}$$

Since water has a density of approximately 1 g/ml (1 kg/L), adding 1 kg of water to 29.8 g of NaCl would produce the desired solution, albeit in a slightly excessive volume (a bit more than 1 L).

Example 5-11:

Calculate the molality of one liter of a seawater solution made from the addition of 29.8 g of NaCl to water to produce a total volume of 1 L.

In this case, you would need the density of seawater (1.025 g/mL) since you are provided with a volume of solution, not a mass of solute.

$$\frac{29.8 \; \cancel{\text{g}} \; \text{NaCl}}{1 \; \cancel{\text{L}} \; \text{solution}} \times \frac{1 \; \text{mol NaCl}}{58.5 \; \cancel{\text{g}} \; \text{NaCl}} \times \frac{\cancel{\text{L}} \; \text{solution}}{1.025 \; \text{Kg solution}} = 0.50 \; \text{m}$$

In Example 5-11, you can see the difference in molality obtained by using volume of solution rather than volume of solvent. In Example 5-10, we made slightly more (about 1 percent more) solution than called for since we weighed the water (as appropriate) and added it to the solute. In the second example, we used the same mass as in Part 1, but it was dissolved in exactly 1 L of solution. The difference in concentration here is 0.01 m, very small but possibly significant, depending on your application.

A shortcut?

What if you didn't have (or couldn't find) the density of the seawater solution? What error would be imparted by using the volume of the solvent (not the solution) only?

Example 5-12:

What would be the error if the volume of solvent—not the volume of solution—were used in Example 5-11?

$$\frac{29.8 \cancel{g} \ NaCl}{1 \cancel{L} \ solution} \times \frac{1 \ mol \ NaCl}{58.5 \cancel{g} \ NaCl} \times \frac{1.0 \cancel{L} \ solution}{Kg \ solution} = 0.51 \ m$$

This amounts to about a 1.9 percent error in solution concentration. This error may be negligible or significant, depending on your application. In general, the more dilute the solution, the less the error in making the assumption that solution density equals solvent density. At infinite dilution, the concentration of the solute is essentially zero (this is an ideal solution). This seawater solution is fairly concentrated. A solution at one-tenth the concentration would have about a 0.2 percent error if the density of the solvent were substituted for the density of the solution. Again, in the laboratory, you will often need to use your judgment to determine the best approach.

SECTION F

ppm, ppb, and Those Other pp's

BACKGROUND AND THEORY

Another concentration unit (commonly found in environmental data) is the ppm or the ppb. Short for parts per million (ppm) and parts per billion (ppb), these simple units confound students. In actuality, the units are quite simple. First up is ppm.

$$ppm = 1 \ part/million \ parts \ or \ 1/10^6$$

Parts per million is a unitless quantity. Since the units of the numerator and denominator are the same, they cancel.

$$\frac{1 \ g}{1,000,000 \ g} \implies \frac{1 \cancel{g}}{10^6 \cancel{g}} = 1 \ ppm$$

If you have one g of solute in 10^6 g of solution, you have a 1 ppm solution. Those of you who despise units may rejoice; however, your joy is short-lived because to make sure you are correctly calculating ppm, you need to keep the units until they cancel. Units strike again!

Similarly, a ppb is 1 part/billion parts, or $1/10^9$

$$\frac{1 \text{ g}}{1,000,000,000 \text{ g}} \Rightarrow \frac{1 \cancel{\text{g}}}{10^9 \cancel{\text{g}}} = 1 \text{ ppb}$$

You can calculate ppt (parts per trillion) and so on. One problem with ppt is that the same name can be used for parts per thousand. If the context is not clear, a worker can be left to guess whether the concentration is 1/1,000 or 1/1,000,000,000,000. Obviously, there is a big difference between the two concentrations. The most common "parts per" values are ppm, ppb, and ppt. Correctly identified parts per thousand should be symbolized as ‰. Parts per hundred is symbolized as ‰, which is our common percentage, discussed in Section G.

In the Laboratory

A little twist to ppm, ppb, and ppt

Since ppm, ppb, and ppt solutions are often very dilute, and in the biological/biochemical laboratory we most often use aqueous solutions, the assumption is often made that the density of the solution is equal to the density of pure water, 1.0 g/ml. Therefore:

$$\frac{1 \cancel{\text{g}} \text{ solute}}{10^6 \cancel{\text{g}} \text{ solution}} \times \frac{1 \cancel{\text{g}} \text{ solution}}{1 \text{ mL solution}} = 1 \text{ ppm}$$

and a ppm is often used as g solute/10^6 mL of solution, specifically for aqueous solutions.

Since we generally see ppm and its counterparts in environmental applications, using water as the solvent makes some degree of sense. In fact, you will often see ppm defined on some websites as g/10^6 mL or g/1000 L. This is not strictly correct, but you can see why the weight volume equivalence is often used in this manner. The parts-per-million, ppm, unit is found routinely in discussing soil concentrations. In this case, it is often expressed as mg substance/kg soil. The difference between the numerator and denominator is appropriately 10^6 (10^{-3} g/10^3 g).

Example 5-13:

As a standard for an environmental analysis, 100 mL of a 4.5 ppm solution of arsenic chloride ($AsCl_3$) is to be prepared.

In this case, all that is needed is the mass and density of the solution. Since 4.5 ppm is very dilute and the significant figures shown indicate a lower level of accuracy needed, the calculation can be accomplished as follows, assuming a density of 1 g/mL for water:

$$4.5 \text{ ppm} = \frac{4.5 \text{ g AsCl}_3}{10^6 \text{ g solution}} \times \frac{1 \text{ g solution}}{1 \text{ mL solution}} \times 100 \text{ mL solution} = 4.5 \times 10^{-5} \text{ g}$$

In this case, the analyst needs to weigh 4.5×10^{-5} g of the arsenic compound and dilute to 100 mL.

TRAP ALERT!

Reality Check:

If you are thinking, you should recognize that the analyst might have a problem with this task. How could you weigh 4.5×10^{-5} g on a standard laboratory balance? Remember, 4.5×10^{-5} g is 0.000045 g, and most laboratory balances weigh to the third or fourth, or some expensive balances weigh to the fifth decimal place. Even if you have one of those fancy balances that weigh to the hundred thousandth place, your uncertainty of measurement would be in the hundred thousandth place—and your measurement would be basically worthless.

So, how could you make this solution? You have two options.

The first is to make more solution. When ingredients are cheap, and often more importantly, disposal is easy, this is often the way to go. In this case, making a liter wouldn't help much because you would still need to weigh 0.00045 g (also not feasible) and you would be disposing of 900 mL of toxic arsenic compound. To accurately weigh your sample on a four-place balance, you would want to weigh at least 0.045 g for best results, and that would require 100 L of solution and a huge amount (99.9 L) of toxic waste.

Here, your second option is the best—serial dilution. Serial dilution involves making a small quantity of a more concentrated solution and diluting that systematically in small volumes to obtain the final diluted concentration. Serial dilutions are covered in Chapter 8 B.

Despite some of the challenges presented with making dilute solutions in the laboratory, the calculations are still quite simple.

SECTION G

Percent (w/w and w/v)

BACKGROUND AND THEORY

Percent is commonly used in the laboratory to express concentration—and that is a shame. Why, you ask? Percent is simple; it is just a part per hundred, or 1/100. We are well acquainted with percentages because our currency is based on 100 pennies in a dollar. Percentages are easy. So, why make a fuss?

The problem with percentages is that for solutions, there are two types of percentages, weight/weight (w/w) and weight/volume (w/v). When a concentration is expressed as a percentage, the w/w or w/v notation should neatly follow in parentheses; however, it rarely does. That can be problematic. The ppm and ppb are often dilute solutions where the mass of the solution and volume of the solution can be related or equated (assuming an aqueous solution) by the simple conversion of 1 g/ml, making them equivalent. However, the pph, or percentage, is generally not dilute by its nature, and solutions expressed as percentage solutions usually have densities that differ fairly significantly from unity. You will often not know whether a particular solution expressed as a percent was made by adding the specified mass of solute to a mass of solution or to a volume of solution. And that difference can be large and cause significant errors.

Technically, percentage, like its counterparts ppm, ppb, and ppt, is defined using like units. A 1 percent (w/w) solution is 1 g solute/100 g solution. The units cancel, and a percentage is a unitless quantity. Remember, when considering masses, they are additive—unlike volumes—and the mass of the solution is the mass of the solute plus the mass of the solvent.

$$\frac{1\,g}{100\,g} \Rightarrow \frac{1\,\cancel{g}}{10^2\,\cancel{g}} = 1\,pph = 1\%$$

If a percentage is not specified as w/w or w/v, a student/technician should assume it is a w/w percentage, as that is the correct way to express percentages. However, if specified as a w/v percentage solution, the expression is similar.

$$\frac{1\,\cancel{g}}{100\,\cancel{mL}} \times \frac{1\,\cancel{mL}}{1\,\cancel{g}} = 1\,pph = 1\%$$

The 1mL/1 g used above is simply the density of water. If the solvent is not water (or the density of the aqueous solution is not unity), then the density of the particular solution must be used. The w/w and w/v percentages are not equivalent.

In the Laboratory

Example 5-14:

Calculate the percentage (w/w) concentration of a solution of NaCl made by dissolving 20 g of NaCl in 200 mL of water.

To accomplish this calculation, you need the mass of water. Assuming a density of the solvent water of 1.0 g/mL, the mass is 200 g

$$\frac{1\ g}{mL} \times 200\ mL = 200\ g$$

and the concentration of the solution is:

$$\frac{10\ g\ NaCl}{10\ g\ NaCl + 200\ g\ water} = \frac{10\ g}{210\ g} \times 100 = 4.8\%$$

Reality Check: What error is imparted by assuming a density of 1 g/mL for this solution? A 5 percent (w/w) solution of NaCl has a density of 1.036 g/mL at room temperature. Using this density, the concentration would be 4.6 percent. This may or may not be significant, depending on your application

Example 5-15:

Calculate the percentage (w/v) concentration of the solution in the previous example, assuming 200 mL of solution is made.

The percentage (w/v) concentration would be calculated as follows:

$$\frac{10\ g\ NaCl}{200\ mL\ water} \times \frac{1.0\ ml\ water}{g\ water} = \frac{10\ g}{200\ g} \times 100 = 5.0\%$$

Example 5-16:

The laboratory needs 200 mL of a 10 percent (w/w) solution of NaCl. Calculate how to prepare the solution.

If you need to make exactly 200 mL, you will need the density of a 10 percent NaCl solution because 200 mL is not necessarily 200 g of a 10 percent solution. But in this case (and in most cases), if you make a bit more or less, you will be fine. In this case, if we make 200 g of solution, we will have less than 200 mL because the density is about 1.074 g/mL for a 10 percent solution. This is because the mass is the sum of the solute and solvent. Since the mass of the solute is not small, we will have less than 200 mL of the solvent and therefore less than 200 mL of solution. We can decide to make less or more. Let's make more first.

Let us make 250 g of solution total:

$$10\% = \frac{10 \text{ g NaCl}}{10 \text{ g solution}} \times 250 \text{ g solution} = 25 \text{ g NaCl}$$

To make this solution, weigh 25 g of NaCl into your vessel. Add water until the total mass is 250 g (that would be 225 g of water by difference). If you weigh the solution, you don't have to bother with density.

But to satisfy curiosity, what would your final volume be?

Using the density:

$$\frac{1 \text{ ml}}{1.074 \text{ g}} \times 250 \text{ g} = 232 \text{ mL}$$

If you didn't think about density and the fact that this is a pretty concentrated solution and assumed that 200 mL would be equivalent to 200 g, you would make the solution as follows:

$$10\% = \frac{10 \text{ g NaCl}}{100 \text{ g solution}} \times 200 \text{ g solution} = 20 \text{ g NaCl}$$

So, you would place 20 g NaCl in the flask and add water until the mass was 200 g.

What would the volume be?

$$\frac{1 \text{ ml}}{1.074 \text{ g}} \times 200 \text{ g} = 186 \text{ mL}$$

You would have your 10 percent solution, but would have about 14 mL less than the protocol specified. That might be fine, depending on the application and exactly how much solution you really needed.

You can see that percentages aren't quite as simple as they seem. The lack of clarity regarding weight or volume in the denominator and the need for densities (which are concentration and temperature dependent) in w/v percentages makes molarity a much more attractive option.

SECTION H

Units (the Biological Type)—U and IU

BACKGROUND AND THEORY

The final unit to cover is the *unit*. The unit is not SI (for good reason) and is generally used in describing biological samples like enzymes. A unit (IU or U) is a quantity of a biologically active substance like an enzyme, hormone, or vitamin, required to produce a specific response. If this seems vague, it is because it is. Like equivalents, units are specific to a particular reaction, or a set of reactions, or a particular effect.

The IU, or international unit, comes from the field of pharmacology. It is also a measurement of biological activity, but for the IU, one IU is equal to one milligram.

An enzyme unit (U) is the amount of a particular enzyme that catalyzes the conversion of one micromole of substrate per minute. To further complicate matters, because of the dependency of enzyme activity on temperature and external conditions, for U, the conditions must be specified. Generally, the temperature is 30°C, and the pH and substrate concentration are adjusted to yield the maximum substrate conversion rate. How do you know what conditions were used? Read the fine print (hopefully, the fine print is provided).

In the Laboratory

How do you know which unit, IU or U, you have? It is best to read the bottle. If your concentration is listed in IU, then you know that the quantity listed is equivalent to mg. If you bought 100 units of an enzyme, and listed on the bottle is a number in units/g or U/g, then you aren't dealing in IU; you most likely have U units. In the latter case, you are given the relationship between enzyme activity and mass. It can sometimes be easier to take your units and convert them to a more standard form of measurement if you need to make solutions.

CHAPTER 6
Converting Units

SECTION A

The Process

BACKGROUND AND THEORY

This chapter provides several examples of interconverting units, a task that is often required in the laboratory and unfortunately, one that often causes significant stress! However, I hope that by following the method outlined and these examples, that stress will dissipate, and you will feel confident when faced with the need to make a molar solution when given a stock solution expressed in percentages.

For example, what if you have a protocol calling for a certain molarity solution and you find that solution, but it is listed as a ppm solution? You need to know whether the solution you have in hand is appropriate to use as is or if it needs to be diluted. Another possibility is that the stock solution is too dilute, and therefore you will need to make another solution and abandon the stock. Hopefully, this section will give you the tools to allow you to be able to answer these questions and be able to accomplish your task.

The way to be successful when you need to make one solution from another or interconvert units is to use dimensional analysis or "unit flipping and unit cancellation" as outlined previously. This means that you MUST use units to accomplish your task. Alternatively, you could try and memorize dozens of formulas, but that is a less-than-successful approach, especially if you don't use the formulas every day.

In order to determine equivalent units, you will often need some basic given information. By *given information*, I mean information like formula mass and densities may be required. Often, this type of information is available on the bottle or in the catalog description, but sometimes, you will have to determine it or find the information in a handbook, in the literature, or on the Internet.

Stick to the process, and you can't go astray

The most common unit, molarity, M, has been chosen as the base unit for solution calculations. In the following examples, conversions from molar concentrations to other units *there* and *back again* are addressed with specific examples. The units chosen are the ones most commonly found in the laboratory. The list is not exhaustive; however, the method is exhaustive and can be applied to conversions between any set of units.

Always dissect or break down compound units

For any conversion between units, the first step is always to *break down* compound units.

For example, if a solution is reported as 0.5 M, it should be expressed as 0.5 mol/L so that the appropriate units are available to use in dimensional analysis. Likewise, percent solutions should be broken down appropriately as either g solute/100 g solution or g solute/100 mL solvent (for dilute aqueous solutions). Part-per-million solutions ($1 \text{ g}/10^6 \text{ g}$) and part-per-billion ($1 \text{ g}/10^9 \text{ g}$) solutions should be treated the same.

In the Laboratory

Normality to Molarity—There and Back Again

SECTION B

N to M and Back Again

To convert between normality, N, and molarity, M, you will need to know something about the chemical behavior of the compound you are using. For the most part, normality is only used for

Brønsted-Lowry acids and bases, and determining the number of equivalents is easily done by simply inspecting the molecular formula of the acid or base (Chapter 5, section D).

In general:

$$N \Rightarrow \frac{eq}{L} \times \frac{mol}{eq} = \frac{mol}{L} = M$$

The trick (although a simple one) for N to M and back again conversions is to know the number of equivalents per mole of solute or the value of the mol/eq fraction. Luckily, this is rather simple for acids and bases.

For monoprotic acids like HCl or monobasic bases like NaOH, only one equivalent of acid or base is provided per mole of acid or base, respectively. In this case, normality equals molarity (N = M). For example, for a 0.1 N solution of a monoprotic acid or monobasic base:

$$0.1\ N = \frac{0.1\ eq}{L} \times \frac{1\ mol}{eq} = \frac{0.1\ mol}{L} = 0.1\ M$$

And Back Again M to N

In general:

$$M \Rightarrow \frac{mol}{L} \times \frac{eq}{mol} = \frac{eq}{L} = N$$

For a 0.1 M solution of HCl:

$$0.1\ M = \frac{0.1\ mol}{L} \times \frac{1\ eq}{mol} = \frac{0.1\ eq}{L} = 0.1\ N$$

For a diprotic acid like H_2SO_4, there are two equivalents of H^+ per mole of acid. The same holds for a dibasic base like $Ca(OH)_2$, where there are two equivalents of hydroxide per mole of base.

The calculation is basically the same:

N to M

$$0.1\ N = \frac{0.1\ eq}{L} \times \frac{1\ mol}{2\ eq} = \frac{0.1\ mol}{L} = 0.05\ M$$

And Back Again M to N

$$0.05 \text{ M} = \frac{0.05 \text{ mol}}{\text{L}} \times \frac{2 \text{ eq}}{\text{mol}} = \frac{0.1 \text{ eq}}{\text{L}} = 0.1 \text{ N}$$

For diprotic acids and dibasic bases, the normality is double the molarity, and the molarity is therefore half of the normality. For triprotics and tribasics, the normality is triple the molarity; the molarity is likewise one-third of the normality.

SECTION C

%(w/w) to M and Back Again

Conversion between percent solution concentration and molarity involves the conversion of mass to moles. This is easily accomplished using the formula mass of the solute. If the solution concentration is expressed as a weight/weight percentage, then the density of the solution (or an assumption about the density) will also be needed. This is because in w/w solutions, the denominator is in terms of mass, and in molar solutions, the denominator is in terms of volume. Therefore, density is needed to convert mass of solution to volume of solution.

%(w/w) to Molarity

For a solution listed as a weight/weight (weight solute/weight solution):
In general:

$$\% \Rightarrow \frac{\text{mass (solute)}}{\text{mass (solution)}} \times \frac{\text{moles (solute)}}{\text{mass (solute)}} \times \frac{\text{mass (solution)}}{\text{volume (solution)}} = \frac{\text{moles (solute)}}{\text{volume (solution)}} = \text{M}$$

Remember, the mass of the solution is the sum of the mass of the solute and the mass of the solvent.

The first step in any percentage calculation is always to express the percentage in terms of g of solute per 100 g of solution.
In units:

$$\% \Rightarrow \frac{\text{g (solute)}}{100 \text{ g (solution)}} \times \frac{\text{mol (solute)}}{\text{g (solute)}} \times \frac{\text{g (solution)}}{\text{L (solution)}} = \frac{\text{moles (solute)}}{\text{L (solution)}} = \text{M}$$

or as a formula:

$$\% \times \frac{1}{\text{formula mass (solute)}} \times \text{density (solution)} = M$$

Keeping the units with the numbers (always!) will keep you from making mistakes. For example, in the calculation outlined above, you must use a density of the solution in g/L in order to obtain the correct denominator for molarity. However, you will most often find densities listed in either kg/L or g/mL. If you use the wrong number without keeping units in place, you will have a molarity that is either 1000 times too small or 1000 times too large. This is the danger in simply memorizing a formula rather than using your units to direct your calculation! If you keep your units in place, you will not make that mistake because if you do, the units will not cancel and you will have to look for your mistake. You can convert the density to the desired units separately as above, or you can expand the calculation to include the conversion of density to the correct units as follows:

For a density expressed in kg/L:

$$\% \Rightarrow \frac{\text{g (solute)}}{100 \text{ g (solution)}} \times \frac{\text{mol (solute)}}{\text{mass (solute)}} \times \frac{\text{kg (solute)}}{\text{L (solution)}} \times \frac{1000 \text{ g (solute)}}{\text{kg (solute)}} = \frac{\text{mol (solute)}}{\text{L (solution)}} = M$$

This has you convert the kg/L term to g as of the calculation, correctly multiplying by 1000.

Or for a density expressed in g/mL (more common):

$$\% \Rightarrow \frac{\text{g (solute)}}{100 \text{ g (solution)}} \times \frac{\text{mol (solute)}}{\text{mass (solute)}} \times \frac{\text{g (solute)}}{\text{mL (solution)}} \times \frac{1000 \text{ ml (solution)}}{\text{L (solution)}} = \frac{\text{mol (solute)}}{\text{L (solution)}} = M$$

Example 6-1:

Calculate the molarity of a 10%w/w solution of aqueous ethanol.

To solve this problem, you will need the formula mass of the solute ethanol (46.0 g/mol) and its density (0.980 g/mL). These values can be looked up or for the formula mass, calculated.

$$10\% \Rightarrow \frac{10 \text{ g (solute)}}{100 \text{ g (solution)}} \times \frac{1 \text{ moles (solute)}}{46 \text{ g (solute)}} \times \frac{0.980 \text{ g (solute)}}{\text{mL (solution)}} \times \frac{1000 \text{ ml (solution)}}{\text{L (solution)}}$$

$$= \frac{2.13 \text{ moles (solute)}}{\text{L (solution)}} = 2.13 \text{ M}$$

Reality Check: A 10 percent solution is fairly concentrated. Therefore, a molarity of 2.13 seems reasonable.

If you did not have the density of the solution and made the assumption that the solution had the same density as water, the calculation would be as follows:

$$10\% \Rightarrow \frac{10 \text{ g (solute)}}{100 \text{ g (solution)}} \times \frac{1 \text{ moles (solute)}}{46 \text{ g (solute)}} \times \frac{1 \text{ g (solute)}}{\text{mL (solution)}} \times \frac{1000 \text{ ml (solution)}}{\text{L (solution)}}$$

$$= \frac{2.17 \text{ moles (solute)}}{\text{L (solution)}} = 2.17 \text{ M}$$

The difference between the two values is 0.04 M, a value that may or may not be significant, depending upon your application.

And Back Again M to %(w/w)

In general:

$$M \Rightarrow \frac{\text{moles (solute)}}{\text{volume (solution)}} \times \frac{\text{mass (solute)}}{\text{moles (solute)}} \times \frac{\text{volume (solution)}}{\text{mass (solution)}} = \frac{\text{mass (solute)}}{\text{mass (solution)}} \times 100 = \%$$

In units:

$$M \Rightarrow \frac{\text{mol (solute)}}{\text{L (solution)}} \times \frac{\text{g (solute)}}{\text{mol (solute)}} \times \frac{\text{L (solution)}}{\text{g (solution)}} = \frac{\text{g (solute)}}{\text{g (solution)}} \times 100 = \%$$

Or as a formula:

$$M \times \text{formula mass (solute)} \times \frac{1}{\text{density (solution)}} \times 100 = \%$$

Keeping the units in place, flipping where necessary, and "stringing" the calculation along, followed by unit cancellation, will keep you from making errors in units. Hence, you won't make an error in the calculation.

Notice here in converting from M to (w/w)% that you are dividing by density and multiplying by formula mass as well as by 100.

Example 6-2:

Determine the %(w/w) concentration of a 2 M solution of aqueous ethanol in water.

As in the previous examples, you will need the formula mass of the solute, ethanol (46.0 g/mol), and the density of the solution (0.979 g/mL). You will need to take care with the units of the density as outlined above.

$$M \Rightarrow \frac{2 \text{ mol (solute)}}{\text{L (solution)}} \times \frac{46.0 \text{ g (solute)}}{\text{mol (solute)}} \times \frac{\text{ml (solution)}}{0.981 \text{ g (solution)}} \times \frac{\text{L (solution)}}{1000 \text{ mL (solution)}}$$

$$= \frac{\text{g (solute)}}{\text{g (solution)}} \times 100 = 9.40\%$$

SECTION D

%(w/v) to M and Back Again

Percent (w/v) solution conversions to molarity are a bit simpler than w/w solutions because the density of the solution is not needed. For %(w/v) solutions, the denominator is already in terms of volume of the solution—not in mass terms. Therefore, no mass-to-volume conversion is needed, and the equation simplifies by a term.

In general:

$$\% \Rightarrow \frac{\text{g (solute)}}{100 \text{ mL (solution)}} \times \frac{1000 \text{ mL (solution)}}{\text{L (solution)}} \times \frac{\text{mol (solute)}}{\text{g (solute)}} \times \frac{\text{mol (solute)}}{\text{L (solution)}} \equiv M$$

It is critical that the percentage be expressed as mass per 100 mL solution. Then, you must remember to convert mL to L. If you keep the units in the calculations, you will not forget this task.

Or as a formula:

$$\% \Rightarrow \frac{\text{mass (solute)}}{\text{volume (solution)}} \times \frac{1}{\text{formula mass (solute)}} = M$$

Example 6-3:

Using the ethanol exemplar in Example 6-2, calculate M for a 10 percent w/v aqueous solution of ethanol.

$$\% \Rightarrow \frac{10 \text{ g (solute)}}{100 \text{ mL (solution)}} \times \frac{1000 \text{ mL (solution)}}{\text{L (solution)}} \times \frac{1 \text{ mol (solute)}}{46.0 \text{ g (solute)}} \times \frac{2.17 \text{ mol (solute)}}{\text{L (solution)}} \equiv 2.17 \text{ M}$$

In this case, a 10 percent (w/v) solution, the molarity is 2.17 M, unlike in the 10 percent (w/w) solution (Example 6-1), where the molarity is 2.13 M. Here again, you can see the problem with percentages if (w/w) or (w/v) is not specified.

And Back Again M to %(w/v)

Conversion of %(w/v) solutions to molar terms is the reverse of the calculation above.
  In general:

$$M \Rightarrow \frac{\text{mol (solute)}}{\text{L (solution)}} \times \frac{1 \text{ L (solution)}}{1000 \text{ mL (solution)}} \times \frac{\text{g (solute)}}{\text{mol (solute)}} \times 100 = \%$$

Or as a formula:

$$M \Rightarrow \frac{\text{moles (solute)}}{\text{volume (solution)}} \times \text{formula mass (solute)} \times 100 = \%$$

To move from M to %(w/v), multiply by formula mass and 100 to obtain the percentage, but mind how the volume is expressed. You must make sure that the mL of solution are correctly converted to L (since molarity is defined in L). If you do not use your units and simply follow the equation above, you will obtain a solution concentration that is too dilute.

Example 6-4:

For a 2M solution of aqueous ethanol, calculate the w/v percentage concentration.

$$M \Rightarrow \frac{2 \text{ mol (solute)}}{\text{L (solution)}} \times \frac{1 \text{ L (solution)}}{1000 \text{ mL (solution)}} \times \frac{46.0 \text{ g (solute)}}{\text{mol (solute)}} \times 100 = 9.20\%$$

Using this example and comparing to the %(w/w) conversion in Example 6-2, we see that there is a 0.2 percent difference in solution concentration between the w/w and w/v percentages.

SECTION E

ppm(b, t) to M and Back Again

This section is purposefully right after percentages to illustrate the parallels between percentage (or parts per hundred), ppm (parts per million), ppb (parts per billion), and ppt (parts per trillion). In percentage calculations, the percent is expressed in like units (g/g), with a difference of 100 between the numerator and denominator. In other words, 10 percent is 10 g/100 g. Likewise, ppm is expressed in like units (g/g), with a difference of 1,000,000 (or 10^6) between the numerator and denominator. The same logic follows for ppb (difference of 10^9) and ppt (difference of 10^{12}). The calculations presented in this section parallel those in the percentage-to-molarity-and-back-again sections.

With ppm, ppb, ppt, and so on the same problem applies with respect to w/w or w/v specificity. Parts per million is defined (default) as g of solute/10^6 g of solution. In the case of ppm, or the even more dilute units for ppb or ppt, for aqueous solutions, there is usually less error involved in assuming that the density of the solution is the same as the mass of the solution. The more dilute the solution, the closer the density of the solution is to the density of the pure solvent. For aqueous solutions, this can generally assumed to be 1 g/mL.

In general:

$$\text{ppm(b,t)} \Rightarrow \frac{\text{mass (solute)}}{\text{mass (solution)}} \times \frac{\text{moles (solute)}}{\text{mass (solute)}} \times \frac{\text{mass (solution)}}{\text{volume (solution)}} = \frac{\text{moles (solute)}}{\text{volume (solution)}} = M$$

Or as a formula:

$$\text{ppm(b,t)} \times \frac{1}{\text{formula mass (solute)}} \times \text{density (solution)} = M$$

The trick to these calculations is making sure you express the ppm correctly (the correct value for the mass [solute]/mass [solution] term above). This is usually where students go astray. The units chosen must differ by 10^6. The simplest ratio to use is g/10^6 g, or mL/10^6 mL. However, you have other options for mass ratios like mg/Kg or μg/g and for volume ratio units like mL/1000 L or μL/L. Each of these options provides a difference of one million, or 10^6 between the units of the numerator and denominator. The decimal point is simply being moved to the right or left as appropriate. You can use the alternate expressions, or use the simple g/g and convert to milli- or micro-values as needed by canceling units.

$$\text{ppm} \Rightarrow \frac{\text{g (solute)}}{10^6 \text{g (solution)}} \times \frac{\text{mol (solute)}}{\text{g (solute)}} \times \frac{\text{g (solute)}}{\text{mL (solution)}} \times \frac{1000 \text{ mL (solution)}}{\text{L (solution)}}$$

$$= \frac{\text{mol (solute)}}{\text{L (solution)}} = M$$

This is similar to the expression for %(w/w) to M, except for the denominator (substituting a million for a hundred). You MUST pay attention to the units used for density so that your final units for molarity are correctly expressed as moles/L. Keeping your units in the calculation will ensure success.

For ppb and ppt, the analogies above hold; however, the difference between numerator and denominator changes to 10^9 or 10^{12}, respectively.

$$\text{ppm} = \frac{\text{g (solute)}}{10^6 \text{g (solution)}} \qquad \text{ppb} = \frac{\text{g (solute)}}{10^9 \text{g (solution)}} \qquad \text{ppt} = \frac{\text{g (solute)}}{10^{12} \text{g (solution)}}$$

Very often, the density in ppm, ppb, and ppt calculations is neglected for aqueous solutions (assumed a value of 1 g/mL), for reasons previously discussed. Therefore, the $\text{g}/10^6 \text{g}$ is commonly expressed in volume terms directly.

In general:

$$\text{ppm(b,t)} \times \frac{1}{\text{formula mass (solution)}} = \text{M}$$

In units for ppm:

$$\text{ppm} \Rightarrow \frac{\text{g (solute)}}{10^6 \text{mL (solution)}} \times \frac{\text{mol (solute)}}{\text{g (solute)}} \times \frac{1000 \text{ mL (solution)}}{\text{L (solution)}} = \frac{\text{mol (solute)}}{\text{L (solution)}} = \text{M}$$

TRAP ALERT! The danger here is that students don't often realize they are converting mass to volume silently, using a density of 1 g/ml. This can be problematic if accurate results are needed and the density of the solution is not unity. In that case, the value obtained for molarity will differ due to the differences in density between pure water and the solution under study. Whether that difference (often small) is significant depends on the application. This parallels the discussion above for w/w and w/v percentages.

Example 6-5:

Determine the molarity of a 10 ppm aqueous solution of $Pb(NO_3)_2$.

To accomplish this task, you will need the formula mass of $Pb(NO_3)_2$ (331.2 g/mol) and the density of the solution (or the assumption of a density of the pure solvent; in this case, 1.0 g/mL is used as an approximation)

$$ppm \Rightarrow \frac{10 \text{ g (solute)}}{10^6 \text{ g (solution)}} \times \frac{1 \text{ mol (solute)}}{331.2 \text{ g}} \times \frac{1.0 \text{ g (solute)}}{\text{mL (solution)}} \times \frac{1000 \text{ mL (solution)}}{\text{L (solution)}}$$

$$= \frac{2.3 \times 10^{-5} \text{ mol (solute)}}{\text{L (solution)}} = 3.02 \times 10^{-5} \text{ M}$$

Reality Check: The molarity found is about 10^{-5} M; a molarity of 10^{-6} would be a 1,000,000 fold or ppm difference. $10 \times 10^{-6} = 10^{-5}$. Therefore, a molarity near 10^{-5} is in the right order of magnitude for a 10 ppm solution.

Example 6-6:

Simplify the calculation in Example 6-5 by expressing the 10ppm solution as g/ml and calculate the molarity:

$$ppm \Rightarrow \frac{10 \text{ g (solute)}}{10^6 \text{ mL (solution)}} \times \frac{1 \text{ mol (solute)}}{331.2 \text{ g}} \times \frac{1000 \text{ mL (solution)}}{\text{L (solution)}}$$

$$= \frac{3.02 \times 10^{-5} \text{ mol (solute)}}{\text{L (solution)}} = 3.02 \times 10^{-5} \text{ M}$$

or more simply by expressing the volume of 10^6 mL as 10^3 L:

$$ppm \Rightarrow \frac{10 \text{ g (solute)}}{10^3 \text{ L (solution)}} \times \frac{1 \text{ mol (solute)}}{331.2 \text{ g}} = \frac{3.02 \times 10^{-5} \text{ mol (solute)}}{\text{L (solution)}} = 3.02 \times 10^{-5} \text{ M}$$

As you can see by Examples 6.5 and 6.6, there is significant flexibility in approaching ppm calculations. All three approaches may be valid. If the solution is aqueous and the density is 1.0 g/mL, the same answers are produced. But if the density differs from unity (the aqueous solution is either not at room temperature or the degree of accuracy required is higher than two or three significant figures), then the w/w and w/v expressions of ppt are not the same. For non-aqueous solutions, the mass volume equivalency cannot be assumed, as the density would not be 1.0 g/mL. However, in any case, you must express the ppm (or ppb or ppt) initial term correctly. You need the correct magnitude of difference between the numerator and denominator.

And Back Again, M to ppb

In general:

$$M \Rightarrow \frac{\text{moles (solute)}}{\text{volume (solution)}} \times \frac{\text{mass (solute)}}{\text{moles (solute)}} \times \frac{\text{volume (solution)}}{\text{mass (solution)}}$$

$$= \frac{\text{mass (solute)}}{\text{mass (solution)}} \times 10^6 = \text{ppm}$$

In units:

$$M \Rightarrow \frac{\text{mol (solute)}}{\text{L (solution)}} \times \frac{\text{g (solute)}}{\text{mol (solute)}} \times \frac{\text{L (solution)}}{\text{g (solution)}} = \frac{\text{g (solute)}}{\text{g (solution)}} \times 10^6 = \text{ppm}$$

Or as a formula:

$$M \times \text{formula mass (solute)} \times \frac{1}{\text{density (solution)}} \times 10^6 = \text{ppm}$$

Notice here in converting from M to ppm that you are dividing by density and multiplying by formula mass as well as by a million. When converting from ppm, you multiply by density and divide by formula mass and by 10^6. Keeping the units in place and stringing the calculation using dimensional analysis will keep you from making errors in units.

Again, the assumption of a density of 1 g/mL will simplify the calculation. But you must be particularly careful with units to make sure that you maintain the correct difference of 10^6 between numerator and denominator. In the expression above, density is expressed in L/g—not mL/g—so our assumption of a density of 1.0 g/mL (correct for dilute aqueous solutions) is really 1000 g/L, or 1 L/1000 g.

This changes the calculation as follows:

$$M \Rightarrow \frac{\text{mol (solute)}}{\text{L (solution)}} \times \frac{\text{g (solute)}}{\text{mol (solute)}} \times \frac{\text{mL (solution)}}{\text{g (solution)}} \times \frac{1 \text{ L (solution)}}{1000 \text{ mL (solution)}}$$

$$= \frac{\text{g (solute)}}{\text{g (solution)}} \times 10^6 = \text{ppm}$$

Alternatively, the expression can be simplified by expressing the molarity as moles per 1000 mL:

$$M \Rightarrow \frac{\text{mol (solute)}}{1000 \text{ mL (solution)}} \times \frac{\text{g (solute)}}{\text{mol (solute)}} \times \frac{\text{mL (solution)}}{\text{g (solution)}} = \frac{\text{g (solute)}}{\text{g (solution)}} \times 10^6 = \text{ppm}$$

It is absolutely critical in ppm calculations to make sure that units are carried along with the calculations. You should see in this example that without units, you will surely make a mistake in using the correct value of density because you will cancel the volume of solution in L with mL for the density and will obtain a solution a thousandfold more concentrated.

Example 6-7:

Calculate the concentration in ppm of a 3.2×10^{-10} M aqueous solution of mercury metal.
To accomplish this problem, the formula mass of mercury metal (Hg, 200.59 g/mol) is required. You must also assume a density of 1.0 g/mL or look up the density of the solution.

$$M \Rightarrow \frac{3.2 \times 10^{-10} \text{ mol (solute)}}{\text{L (solution)}} \times \frac{200.59 \text{ g (solute)}}{\text{mol (solute)}} \times \frac{1 \text{ mL (solution)}}{1 \text{ g (solution)}} \times \frac{1 \text{ L (solution)}}{100 \text{ mL (solution)}}$$

$$= \frac{6.42 \times 10^{-11} \text{ g (solute)}}{\text{g (solution)}} \times 10^6 = 6.42 \times 10^{-5} \text{ ppm}$$

Reality Check: This is an extremely dilute solution, as expected from such a small molarity. In this case, it makes more sense to answer the question in terms of ppb or even ppt:

$$6.42 \times 10^{-5} \text{ ppm} \times \frac{1000 \text{ ppb}}{\text{ppm}} = 6.42 \times 10^{-2} \text{ ppb}$$

Or:

$$6.42 \times 10^{-5} \text{ ppm} \times \frac{10^6 \text{ ppt}}{\text{ppm}} = 6.42 \text{ ppt}$$

CHAPTER 7
From Solution to Solution

SECTION A

The Dilution Equation: Your Best Friend in the Laboratory

BACKGROUND AND THEORY

I f you need to make a particular solution and you have on hand a more concentrated solution of the same composition, you are in luck. Simple dilution is the answer! Even better, there is a simple equation that will let you know exactly what to do. This equation is called the dilution equation and can be stated as:

$$C_1V_1 = C_2V_2$$

where C_1 refers to the concentration of the stock solution (the more concentrated solution), and V_1 refers to its volume. C_2 refers to the concentration of the desired solution (the more dilute solution), and V_2 refers to the desired volume of solution C_2.

Correctly using the dilution equation is the key

To correctly use the dilution equation, the units of C_1 and C_2 must be identical! The same holds for V_1 and V_2. In fact, the dilution equation is a simple proportion that has been cross multiplied. The concentration units and the volume units can be any units, as long as they are the same. Why this works:

$$\frac{mol}{L} \times L = \frac{mol}{L} \times L$$

Above the dilution equation is expressed using SI units. You can easily see that this equation is true. It reduces to $1 = 1$.

$$\frac{mol}{L} \times L = \frac{mol}{L} \times L \Rightarrow M \times L = M \times L \Rightarrow 1 = 1$$

But the following is also true:

$$\frac{mol}{L} \times mL = \frac{mol}{L} \times mL \quad OR \quad \frac{mol}{Kg} \times L = \frac{mol}{Kg} \times L$$

You can substitute any concentration units for concentration, C, or any volume units for volume, V, as long as the units for both the stock and diluted solution concentrations are the same and the units for both volumes are the same. In the first equation, the SI units, molarity and liters, are used. In the subsequent equation, molarity is used with mL, and molality is then used with L. All of these are valid since they all reduce to the same true expression, $1 = 1$.

The ability to use any concentration units combined with any volume is incredibly convenient. Often, the dilution equation is taught as $M_1V_1 = M_2V_2$ in freshman chemistry. This is most likely because molarity is the most commonly used solution unit. However, students don't realize that this equation is just a simple proportion and that any units can be used.

In the Laboratory

The dilution equation is one of the most useful in the laboratory. It is especially useful if you need to make several solutions of the same compound of different concentrations. In this case, you can make (or obtain) a stock solution and then make all of your desired solutions via dilution of the stock. This is much simpler and often more accurate than making several solutions independently. The dilution equation will allow you to easily calculate how much (V_1) of the stock (C_1) that you will need to make enough (V_2) of the desired concentration (C_2). Generally, V_1 is the variable since you most often know the concentration of the stock and you know the volume and concentration you need of the diluted solution.

Example 7-1:

You need to make 100 mL of a 100 mM solution of aqueous dibasic sodium phosphate. In the lab, you find a 0.5 M solution.

In this instance, you need to know how much of the 0.5 M solution you need. The unknown is V_1, the volume of the stock will you need.

$$C_1 = 0.5 \text{ M}, C_2 = 100 \text{ mM and } V_2 = 100 \text{ mL}$$

Before you can use the dilution equation, you MUST make the units of C_1 and C_2 agree. You can choose to use either M or mM—it doesn't matter, as long as they are the same. Using mM:

$$0.5 \text{ M} \Rightarrow \frac{0.5 \text{ mol}}{L} \times \frac{1000 \text{ mmol}}{\text{mol}} = 500 \frac{\text{mmol}}{L} = 500 \text{ mM}$$

and

$$C_1 V_1 = C_2 V_2$$

substituting values:

$$500 \text{ mM} \times V_1 = 100 \text{ mM} \times 100 \text{ mL}$$

and

$$V_1 = 20 \text{ mL}$$

To make the 100 mM solution, take 20 mL of the stock solution, and dilute to 100 mL.

Those units again!

As always, use your units! Keeping the units in the calculation will keep you from forgetting to make sure the units are consistent, and that will make the numbers correct. Also, keeping the units in the equation gives you the units for your solution, so you know what volume to measure.

Determining Stock Concentration

You can also use the dilution equation backward. In other words, you can use the dilution equation to determine a good concentration of stock to make. This often confuses students because there is not one correct answer. As long as your choice works (that is, gives you the desired concentrations and is possible to make with the materials on hand), it is correct.

Example 7-2:

Prepare a stock that can be used to make five 100 mL solutions of NaCl in molarities ranging from 0.001 to 0.5 M.

In a problem like this, you need to make several NaCl solutions. This is common when performing an assay when standards are required. The stock solution must be in greater concentration than the most concentrated solution. In addition, you want to make sufficient quantities. With a compound like NaCl, waste isn't really an issue, since salt is cheap and excess can be washed down the drain without harming any fish! However, if your solution is expensive or toxic, you will want to carefully plan to minimize waste. In this case, use the most concentrated desired solution to calculate an appropriate volume of the stock. All other dilutions can be made from the same stock. Here, 0.5 M is the most concentrated solution. This is C_2. The needed volume is 100 mL, V_2. If you choose a convenient volume to pipette, say, 10 mL, you can assign this to V_1. (***Note:*** This choice is arbitrary, and a different convenient volume can be chosen.)

$$C_1V_1 = C_2V_2$$

substituting values:

$$C_1 \times 10 \text{ mL} = 0.5 \text{ M} \times 100 \text{ mL}$$

and

$$C_1 = 5M$$

A 5 M stock of NaCl could be used. To calculate how to make a 5 M solution, you follow the same steps as for making any molar solution. You need the formula mass of NaCl (58.5 g/mol), and you also need to decide how much of the stock to make. Since your goal is 10 mL for the first solution (the most you will need for any solution) and you have four solutions remaining, 50 mL of solution will provide sufficient stock. If you want some to store (and since NaCl is cheap), you could make 100 mL and have plenty left over for mistakes or spillage (or repeating an experiment!).

$$\frac{0.5 \text{ mol}}{L} \times \frac{1 \text{ L}}{1000 \text{ mL}} \times \frac{58.5 \text{ g}}{mol} \times 100 \text{ mL} = 2.93 \text{ g}$$

Weigh 2.93 g of NaCl and dilute to 100 mL with distilled water.

Remember to check solubility

You must always check solubility and make sure that your stock is soluble at the concentration you have calculated. If it is not soluble, you need to make a more dilute stock solution. Therefore, adjust your volume, V_1, upward so that a less concentrated solution is needed—and try again!

For your other solutions, you would now use this 0.5 M stock to make your solutions. But what do you do if V_1, the volume of the stock needed, is too small to measure accurately? You become a serial dilutionist!

SECTION B
Serial Dilutions—Making Very Dilute Solutions

BACKGROUND AND THEORY

What if you solve the dilution equation and find that you need to dilute 0.1 µL of your stock to obtain your final solution? You simply cannot accurately measure volumes this small in most laboratories.

If you need to make a particularly dilute solution or if your stock solution is too concentrated, you may find that the volume of the stock you require to make the diluted solution is too small to be measured accurately. In this instance, you must use a process called serial dilution.

In **serial dilution**, one or more intermediate solutions is made from the original stock. The final desired solution is made from the intermediate solution or solutions.

In the Laboratory

There is no single way to do a serial dilution correctly. The concentrations and volumes of any intermediate solutions may vary from worker to worker. It is up to you to choose, based upon your particular task, what volume or volumes of intermediates and what concentration or concentrations to make. As long as the final concentration of the desired solution is correct, you have successfully completed your task. Of course, you should always pay attention to the quantity of reagents used and the waste generated. There is no need to make 1 liter of solution, for example, if you will only need 10 mL.

Example 7-3:

You need to make 100 mL of a 0.10 mM solution of aqueous dibasic sodium phosphate. In the lab, you find a 1 M solution.

Begin by using the dilution equation. One concentration is molar and the second mM. Both concentrations must be expressed in the same units. Choosing to convert M to mM:

$$1\text{ M} = \frac{1\text{ mol}}{L} \times \frac{1000\text{ mmol}}{mol} = \frac{1000\text{ mmol}}{L} = 1000\text{ mM}$$

knowing:

$$C_1V_1 = C_2V_2$$

substituting values:

$$1000\text{ mM} \times V_1 = 0.10\text{ mM} \times 100\text{ mL}$$

leads to:

$$V_1 = 0.001\text{ mL or 1 }\mu\text{L}$$

To make this solution, you would need 1 μL of your stock solution diluted to 100 mL. However, 1 μL is generally near the lower limit of benchtop pipettors. Also, the accuracy of such a small volume (as a percentage of the total volume) can be large. In this case, you are probably better off using a larger volume of stock.

There are two approaches here. You can simply make more than 100 mL of the final solution. For instance, if you choose to make 1 L, you would need 10 μL of the stock. However, you would waste 900 mL of your solution.

The second approach involves using serial dilution. Instead of making the final desired solution directly from the stock, an intermediate concentration solution can be made and then the final solution prepared from that intermediate solution. The choice of intermediate concentration varies, but it is often convenient to use increments of ten, since simple movement of decimal places can provide changes in concentrations.

In this instance, if the original stock solution was diluted a hundredfold by taking 1 mL of that stock and diluting to 100 mL, the intermediate stock concentration calculated using the dilution equation would be:

$$C_1V_1 = C_2V_2$$

substituting values:

$$1000 \text{ mM} \times 1 \text{ mL} = C_2 \times 100 \text{ mL}$$

and

$$C_2 = 10 \text{ mM}$$

The desired 100 mL of the final solution at 0.10 mM could next be made, also using the dilution equation:

$$C_1 V_1 = C_2 V_2$$

substituting values:

$$10 \text{ mM} \times V_1 = 0.1 \text{ mM} \times 100 \text{ mL}$$

leads to:

$$V_1 = 1 \text{ mL}$$

To accomplish your task, prepare an intermediate solution by pipetting 1 mL of the 1 M and diluting to 100 mL with distilled water. Then, pipette 1 mL of the newly made solution into a second flask and dilute to 100 mL with distilled water. The final concentration will be 0.1 mM, the desired solution concentration.

Another way to do the same thing

Example 7-4:

But what would happen if we chose a tenfold dilution instead of our 100 fold dilution in Example 7-2?

For instance, if we first took 10 mL of the stock and diluted to 100 mL:

$$C_1 V_1 = C_2 V_2$$

substituting values:

$$1000 \text{ mM} \times 10 \text{ mL} = C_2 \times 100 \text{ mL}$$

leads to:

$$C_2 = 100 \text{ mM}$$

The desired 100 mL of the final solution at 0.10 mM could next be made:

$$C_1 V_1 = C_2 V_2$$

substituting values:

$$100 \text{ mM} \times V_1 = 0.1 \text{ mM} \times 100 \text{ mL}$$

leads to:

$$V_1 = 0.1 \text{ mL}$$

In this case, the intermediate would be made by diluting 10 mL of the stock to 100 mL and the final solution made by taking 100 μL (0.1 mL) of the intermediate and diluting to 100 mL. Either way, the final concentration is correct. In both cases, the volumes were measurable.

You could abandon the multiples of tens and instead choose any volume if you wanted. For example, what if you took 5 mL of the stock and diluted it to 250 mL?

$$C_1 V_1 = C_2 V_2$$

substituting values:

$$1000 \text{ mM} \times 5 \text{ mL} = C_2 \times 250 \text{ mL}$$

leads to:

$$C_2 = 20 \text{ mM}$$

The desired 100 mL of the final solution at 0.10 mM could next be made:

$$C_1 V_1 = C_2 V_2$$

substituting values:

$$20 \text{ mM} \times V_1 = 0.1 \text{ mM} \times 100 \text{ mL}$$

leads to:

$$V_1 = 0.5 \text{ mL}$$

In this respect, the intermediate is made by diluting 5 mL of the stock to 250 mL and the final solution made by taking 500 µL (0.5 mL) of the intermediate and diluting to 100 mL. The final concentration is also correct.

Hopefully, this example has shown you that the concentrations of the intermediates should be chosen for convenience and that there is more than one correct way to accomplish the task.

Multiple dilutions

For very concentrated stocks or very dilute solutions, you may need more than one intermediate dilution.

Example 7-5:

A protocol calls for making 1 mL of a 1 nM (nanomolar) solution of a particular dinucleotide triphosphate. The dinucleotide triphosphate is sold as a 0.1 M solution.

There is a 10^8 concentration difference between the stock and the desired solution. With such a large difference and a fairly expensive stock solution, you want to minimize waste. Here, two or more dilutions would make sense.

One way to accomplish this task would be to dilute by a constant amount, say, a thousandfold at each step until you reach the final step. You would want to use a small quantity of the stock solution.

So, you could take 0.1 mL of the stock and dilute to 100 mL (a 10^3 difference).

$$C_1 V_1 = C_2 V_2$$

substituting values:

$$0.1 \text{ M} \times 0.1 \text{ mL} = C_2 \times 100 \text{ mL}$$

leads to:

$$C_2 = 0.0001 \text{ M} = 0.1 \text{ mM}$$

You could then repeat the dilution using your intermediate:

$$C_1 V_1 = C_2 V_2$$

substituting values:

$$0.1 \text{ mM} \times 0.1 \text{ mL} = C_2 \times 100 \text{ mL}$$

leads to:

$$C_2 = 0.0001 \text{ mM} = 0.1 \text{ } \mu\text{M}$$

The desired solution is 1 nM, and a 0.1 μM solution is 100 times more dilute. To make the final solution from the second intermediate, you must reconcile the concentration units (you cannot use the dilution equation with μM and nM as C_1 and C_2).

$$1 \text{ nM} = \frac{1 \text{ nmol}}{\text{L}} \times \frac{1 \text{ } \mu\text{mol}}{1000 \text{ nmol}} = \frac{0.001 \text{ } \mu\text{mol}}{\text{L}} = 0.001 \text{ } \mu\text{M}$$

Using the dilution equation

$$C_1 V_1 = C_2 V_2$$

substituting values:

$$0.1 \text{ } \mu\text{M} \times V_1 = 0.001 \text{ } \mu\text{M} \times 100 \text{ mL}$$

and

$$V_1 = 1 \text{ mL}$$

For the final dilution, take 1 mL of the second intermediate and dilute to 100 mL.

Many different combinations of concentrations and volumes could be chosen at each step. As long as the final concentration is correct, the method is correct.

SECTION C

The Dilution Factor

The dilution factor often confuses students, and it shouldn't. The dilution factor is just the fractional expression of ratio of the initial volume (V_1) to the final volume (V_2). This is easiest to see by looking at an example.

If 1 mL of a sample is diluted to 100 mL, then the ratio of dilution is:

$$V_1{:}V_2$$

or

$$1{:}100$$

The fractional expression is

$$V_1/V_2$$

or in this case

$$1/100, \text{ which could also be expressed as } 0.01.$$

The final solution is one one-hundredth as dilute as the stock solution.

CHAPTER 8
Chemistry Again:Equilibra

SECTION A

What's Temperature Got to Do with It?

BACKGROUND AND THEORY

In Chapter 5, which dealt with solution concentrations, we mentioned the effects of temperature. Temperature affects how molecules move. When heated, they move faster. So, whenever any measurement is made volumetrically (including molarity), there is a temperature effect that must be considered. Often, it is considered and discarded; however, you must be aware of temperature effects because every so often small changes due to changing temperatures lead to large errors.

Volumes of solvents, and thus solutions, are temperature dependent. You know this intuitively. As you heat water, it expands; its volume increases. In order to accurately measure a volume and compare it to another measured volume, the two measurements must be accomplished at the same temperature. One advantage of %(w/w) and molal (m) solutions is that the solution or solvent, respectively, is expressed as mass, not as volume. There is no temperature effect on solution concentration in these terms.

We most often assume STP—standard temperature and pressure—conditions in the laboratory. Standard temperature is 0°C. However, data are more often tabulated and reported at 25°C, which is commonly referred to as room temperature. To be perfectly accurate, a notation of the temperature should always accompany the concentration for any solution. However, this detailed labeling is rarely observed. Generally, this is not problematic as it is for aqueous solutions; the error introduced is small. The thermal coefficient of expansion of water is 0.00021 per 1° Celsius at 20° Celsius. This means that every 1° increase in temperature is accompanied by a very small change in volume. Table 8-1 shows the volume occupied by 1 gram of water as temperature increases.

TABLE 8-1 Volume Occupied by 1 g of Water at
Various Temperatures[1]

Temperature (°C)	Volume (mL)
17.0	1.0022
18.0	1.0024
19.0	1.0026
20.0	1.0028
21.0	1.0030
22.0	1.0033
23.0	1.0035
24.0	1.0037
25.0	1.0040
26.0	1.0043

[1]Perlman, Howard. Densities of water at different temperatures, US Department of the Interior, US Geological Survey (USGS), The Water Science School, 05/01/2015, http://water.usgs.gov/edu/density.html, accessed 05/20/2015.

In the Laboratory

Inspection of the data in Table 8-1 shows a linear dependence of volume on temperature. The expansion of water over nearly a ten-degree temperature range (near room temperature) is less than 0.0025 mL (or 0.25 μL). Most often, the variation in laboratory temperature is at most a few degrees. Considering that we commonly measure volumes with accuracy of +/−0.01 mL or less, the error imparted by not considering temperature is not significant.

The moral of this tale is that you can go on ignoring temperature for the most part. You should, however, be aware that there is a small error introduced if temperatures are not specified and kept constant. Pay attention when considering the degree of accuracy needed for your particular task.

SECTION B

Acids and Bases—the Strong versus the Weak

BACKGROUND AND THEORY

Somewhere in general chemistry, you learned about Brønsted-Lowry acids and bases. Brønsted-Lowry acids donate a proton, and the bases accept a proton. The context of Brønsted-Lowry acids and bases is most often (and most practically for the applications considered in this text) in aqueous solution.

As a reminder, there really is no such thing as H^+, a bare proton, in an aqueous solution. The proton donated in an acid base solution is really "transferred" via the surrounding water. Rather than H^+, the active species is H_3O^+, the hydronium ion. However, you will commonly find the acid referred to as H^+ rather than as H_3O^+. It is understood that in water, H^+ means H_3O^+. Both notations are used in this text.

A typical acid base neutralization reaction, with acid expressed as the proton, and more correctly as hydronium ion, follows:

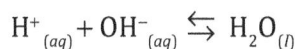

$$H^+_{(aq)} + OH^-_{(aq)} \leftrightarrows H_2O_{(l)}$$

or

$$H_3O^+_{(aq)} + OH^-_{(aq)} \leftrightarrows 2H_2O_{(l)}$$

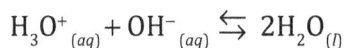

A neutralization reaction involves the combination of acid and base to produce water.

What we are generally concerned with in the laboratory is the concentration of the active acid species, H^+, in solution.

Complete reactions—the strong

For strong acids, we assume that virtually all (100 percent) of the acid has disassociated into ions in solution. Therefore, for a monoprotic acid, the initial concentration of the acid is equal to the final concentration of H^+ in solution.

For the general strong acid HA:

$$HA_{(aq)} \leftrightarrows H^+_{(aq)} + A^-_{(aq)} \quad Or \quad HA_{(aq)} + H_2O_{(l)} \leftrightarrows H_3O^+_{(aq)} + A^-_{(aq)}$$

HA is commonly referred to as the acid, H^+ the proton, H_3O^+ the hydronium ion, and A^- is the conjugate base, often called the salt.

This strong acid reaction proceeds to the right completely (leaving no appreciable concentration of HA). This is often indicated by using only a forward arrow for the reaction, suggesting that there is no reverse or back reaction occurring. You must remember that there is always some reverse reaction; no reaction is ever truly 100 percent complete. However, for practical purposes, the forward reaction for strong acids disassociating in water is so favorable that neglecting the reverse reaction imparts no significant error.

For a strong monobasic base like NaOH, the treatment is parallel to the acid:

$$NaOH_{(aq)} \leftrightarrows Na^+_{(aq)} + OH^-_{(aq)}$$

and the concentration of hydroxide at equilibrium is equal to the initial concentration of sodium hydroxide.

For strong diprotic and triprotic acids, the concentration of the free acid, H^+, is determined using the molar ratio:

$$H_2A_{(aq)} \leftrightarrows 2H^+_{(aq)} + A^{-2}_{(aq)} \quad Or \quad H_2A_{(aq)} + 2H_2O_{(l)} \leftrightarrows 2H_3O^+_{(aq)} + A^{-2}_{(aq)}$$

and

$$H_3A_{(aq)} \leftrightarrows 3H^+_{(aq)} + A^{-3}_{(aq)} \quad Or \quad H_3A_{(aq)} + 3H_2O_{(l)} \leftrightarrows 3H_3O^+_{(aq)} + A^{-3}_{(aq)}$$

for diprotic and triprotic acids, respectively. (See Section D in this chapter for more information on multiprotic acids)

Of course, when using the molar ratio of protons, this assumes that all protons are 100 percent released in water. It is possible—and common—that the first proton is fully disassociated, while the second or third is not. Therefore, some multiprotic acids exhibit the properties of both strong and weak acids.

The parallel treatment for dibasic compounds also holds. In general, metal hydroxides behave as strong bases.

Incomplete reactions—the weak

What if the proton is not 100 percent released? This is the case for weak acids. If only a portion of the protons are released, then the concentration of H^+ at the reaction "end" is not equal to the initial concentration of the acid. The concentration of H^+ is somewhat less than the initial concentration of the acid. Some intact acid (associated with the proton) remains when the reaction is complete. What must be determined here is just how much remains? The proportion or ratio of product (H^+) to reactant (HA) is governed by a value called the equilibrium constant.

The Equilibrium Constant

For any reaction, the equilibrium constant is used to quantify the position of the equilibrium. This is fancy-speak for how far does the reaction proceed to the right (forward reaction) versus to the left (back reaction). The equilibrium constant is a measure of the *balance of the reaction*.

For the general reaction:

$$aA + bB \leftrightarrows cC + dD$$

For liquid solutions, the equilibrium constant is expressed as a ratio of products to reactants and is denoted using a capital K or K_{eq} for equilibrium constant. The coefficients in the balanced equation become exponents in the equilibrium constant:

$$K_{eq} = \frac{[C]^c [D]^d}{[A]^a [B]^b}$$

The brackets represent molar concentrations (M) of all species of reactant and product. This point about units is critical, as units must be M—not millimolar (mM) or micromolar (µM). The equilibrium constant can be expressed using partial pressures for gaseous solutions or in other units. However, for the purpose of solution preparation, in the life sciences, molar rules.

For our monoprotic acid:

$$HA_{(aq)} + H_2O_{(l)} \leftrightarrows H_3O^+_{(aq)} + A^-_{(aq)}$$

and the equilibrium constant is:

$$K_{eq} = K_a = \frac{[H_3O^+]^c [A^-]^d}{[HA]^a [H_2O]^b}$$

Where a, b, c, and d are 1 from the balanced equation. The K_{eq} is called K_a to denote that it is for an acid reaction (see next section for a complete discussion of the K_a). In addition, since water is the solvent, its concentration is considered unchanging. The inclusion of the small amount of water consumed in the reaction is insignificant. Therefore, water is not included in the equilibrium expression. The expression reduces to:

$$K_a = \frac{[H_3O^+][A^-]}{[HA]}$$

For a weak acid, to determine the amount of free acid at equilibrium, the value of the K_a must be obtained and the ratio solved.

Rearranged:

$$[H_3O^+] = \frac{K_a[HA]}{[A^-]}$$

For a monoprotic weak base like NH_3, the treatment is parallel to the weak acid:

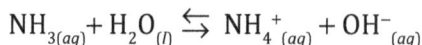

$$NH_{3(aq)} + H_2O_{(l)} \leftrightarrows NH_4^+{}_{(aq)} + OH^-{}_{(aq)}$$

The concentration of hydroxide and the equilibrium constant can be expressed as:

$$K_{eq} = K_b = \frac{[NH_4^+]^c[OH^-]^d}{[NH_3]^a[H_2O]^b}$$

Here, the K_{eq} is called K_b to denote that it is for a base reaction. The concentration of the solvent is removed as before and the exponents are all one, based upon the balanced equation, so the expression reduces to:

$$K_b = \frac{[NH_4^+][OH^-]}{[NH_3]}$$

Similarly, for a weak base, to determine the amount of free base at equilibrium, the value of the K_b must be obtained and the ratio solved.

$$[OH^-] = \frac{[NH_3]K_b}{[NH_4^+]}$$

In the Laboratory

Strong Acids:

The strong acid most often encountered in the laboratory is HCl.

$$HCl_{(aq)} \leftrightarrows H^+{}_{(aq)} + Cl^-{}_{(aq)} \quad \text{Or} \quad HCl_{(aq)} + H_2O_{(l)} \leftrightarrows H_3O^+{}_{(aq)} + Cl^-{}_{(aq)}$$

Since the acid is strong, the reaction is assumed to be complete, and the initial concentration of acid (HCl) is equal to the concentration of hydronium ion in solution at equilibrium. If you initially have a 0.1 M solution of HCl in water, you have a 0.1 M concentration of H^+ at equilibrium (and zero concentration of HCl).

For a strong diprotic acid like H_2SO_4 in water, the concentration of H^+ is twice the concentration of the acid by the balanced equation:

$$H_2SO_{4(aq)} \leftrightarrows 2H^+_{(aq)} + SO_4^{-2}{}_{(aq)} \quad \text{Or} \quad H_2SO_{4(aq)} + 2H_2O_{(l)} \leftrightarrows 2H_3O^+_{(aq)} + SO_4^{-2}{}_{(aq)}$$

Since each mole of H_2SO_4 releases two moles of protons, a 0.1 M solution of H_2SO_4 will produce 0.2 M solution of hydronium ion in water at equilibrium (and zero concentration of acid at equilibrium).

Note: H_2SO_4 is not a typical diprotic acid. The first disassociation is strong; the second is weaker, but still strong. For most di- and triprotic acids, the second and third disassociations are not even close to complete. Therefore, the simple molar ratio cannot be used, and the equilibrium expression must be used as described for weak acids.

For strong bases like NaOH:

$$NaOH_{(aq)} \leftrightarrows OH^-_{(aq)} + Na^+_{(aq)}$$

Since the base is strong, the reaction is assumed to be complete and the initial concentration of base (NaOH) is equal to the concentration of hydroxide ion in solution at equilibrium. If you prepare a 0.1 M solution of NaOH in water, you have a 0.1 M concentration of OH^- at equilibrium.

Weak Acids:

As covered previously, for strong acids, the concentration of hydronium ion at equilibrium is assumed to be equal to the initial concentration of the acid. In other words, all of the acid disassociates to provide protons. Therefore, the concentration of the associated acid, HA, is essentially zero. For weak acids and bases, you need to use the K_a or K_b values and the equilibrium expression to determine the concentrations of acid or base at equilibrium. For a weak acid, at equilibrium, the associated acid, HA, still has appreciable concentration and cannot be neglected; the concentration of the hydronium ion therefore must be somewhat less than the initial concentration of the weak acid HA.

Solving for a weak acid is actually fairly easy. It simply requires solving the equilibrium expression. For our general weak acid:

$$K_a = \frac{[H_3O^+][A^-]}{[HA]}$$

Generally, we are looking for the concentration of hydronium ion. Following the logic and steps outlined below should simplify this calculation.

To solve for the hydronium ion (and presumably the pH), we need some information about the acid. We can look up the K_a value for any weak acid or base on the Web or in indexes. Hydronium ion concentration, $[H_3O^+]$, is an unknown and is assigned the value x. The concentration of the conjugate base $[A^-]$ must be equal to the concentration of hydronium ion based upon the balanced equation:

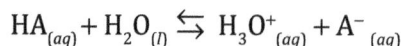

$$HA_{(aq)} + H_2O_{(l)} \rightleftarrows H_3O^+_{(aq)} + A^-_{(aq)}$$

Based upon this equation, for each mole of hydronium ion produced, one mole of anion is produced (this is true for 1:1 relationships). Therefore, $[A^-]$ is also equal to x (if two moles of anion were produced, then the value would be assigned as $2x$). Usually, we know the initial concentration (i) of the intact acid, HA, because that is the solution we are making. However, at equilibrium, some of that acid has disassociated, so the equilibrium concentration is less than the initial concentration. How much less? x less. Hence, the concentration of HA is $i - x$.

Substituting into the equilibrium expression:

$$Ka = \frac{[x][x]}{[i - x]} = \frac{[x^2]}{[i - x]}$$

Now we have one equation and only one variable, so the equation (with a little algebra) can be solved.

Some of my colleagues teach weak acids and bases by a method they call the box method. The box helps keep track of concentrations using a table. If this method works for you, by all means, use it.

The Box Method:

In the box method, a table is constructed to keep track of all reactants and products (listed across the top of the box) and their concentrations at the beginning of the reaction and at equilibrium.

Initially, the concentrations of hydronium and conjugate base are 0 and the concentration of acid is the initial concentration, i.

Concentration	HA	H_3O^+	A^-
Initial	i	0	0
Equilibrium			

At equilibrium, the rest of the box can be filled in:

Concentration	HA	H_3O^+	A^-
Initial	i	0	0
Equilibrium	$i - x$	x	x

At equilibrium, the concentrations of hydronium and conjugate base are x and the concentration of acid had been reduced by x, so it is equal to the initial concentration less the amount disassociated, $i - x$.

Sometimes, it is easier to see how by using actual numbers.

Example 8-1:

What is the hydronium ion concentration of a 0.1 M solution of acetic acid?

The first step in solving this problem is obtaining the correct K_a value for acetic acid. Acetic acid has a K_a of 1.78×10^{-5}. The unknown is hydronium ion concentration, and its value is assigned x (the acetate ion $^-OC(O)CH_3$ is shortened to OAc^- in this example). From the balanced equation:

$$HOAc_{(aq)} + H_2O_{(l)} \rightleftharpoons H_3O^+_{(aq)} + OAc^-_{(aq)}$$

There is a 1:1 molar relationship between the concentrations of hydronium ion and acetate ion (OAc^-), so $[OAc^-]$ is also equal to x. The initial concentration of acetic acid is given as 0.1 M, so at equilibrium, the concentration is reduced by x and can be represented by $0.1 - x$. Our expression becomes:

$$1.78 \times 10^{-5} = \frac{[x^2]}{[0.1 - x]}$$

Using the box method:

Concentration	HOAc	H_3O^+	OAc^-
Initial	0.1	0	0
Equilibrium	$0.1 - x$	x	x

Initially, the concentrations of hydronium and conjugate base are 0, and the concentration of acid is 0.1 M at equilibrium; the values of x and $0.1 - x$ are used.

Our K_a expression is a quadratic equation. We have one variable and one equation, so it is solvable (using the quadratic formula). However, the quadratic formula is one you have most likely forgotten, and there is a trick (assumption) that can be used to further simplify the math. If you make this assumption, you must remember to check your answer and determine whether the assumption is valid.

A little trick to simplify the math

Since the K_a is small (1.78×10^{-5}), very little acid has actually disassociated. In order to obtain a small number, the numerator (hydronium and conjugate base concentrations) must be small and the denominator (acid HA concentration) must be large. A small number divided by a large number yields a small number. Therefore, not much hydronium ion has been produced, and we will make the assumption that the initial concentration of acid is pretty much equal to the equilibrium concentration of acid, or:

$$0.1 - x \approx x$$

With this substitution, our equation becomes:

$$1.78 \times 10^{-5} = \frac{[x^2]}{[0.1]}$$

and

$$x = 1.33 \times 10^{-3} \text{ M}$$

This is much simpler to solve.

Now, you must check the assumption you made to make sure it is true. You assumed that the concentration of HA at equilibrium was essentially the same as it was initially. You assumed that very little reaction occurs. Initially, the acid concentration was 0.1 M. At equilibrium, it is $0.1 - x$, or $0.1 - 1.33 \times 10^{-3}$, which is 0.099M. This is a difference of about 1 percent. Thus, making the assumption introduced about 1 percent error but saved some algebra.

When you have to do the math

Solving a Quadratic Equation:

If 1 percent error is too high, you will have to solve the quadratic formula. If you have forgotten the standard form of a quadratic equation and the quadratic formula, here they are:

$$ax^2 + bx + c = 0$$

and

$$x = \frac{-b \pm \sqrt{b^2 - 4ac}}{2a}$$

Looking at our equation:

$$1.78 \times 10^{-5} = \frac{[x^2]}{[0.1 - x]}$$

We must rearrange it to the quadratic form, so we have to cross multiply.

$1.78 \times 10^{-5} \times [0.1 - x] = [x^2]$

and

$1.78 \times 10^{-6} - 1.78 \times 10^{-5}x = x^2$

and

$x^2 + 1.78 \times 10^{-5}x - 1.78 \times 10^{-6} = 0$

therefore,

$$a = 1, b = 1.78 \times 10^{-5} \text{ and } c = 1.78 \times 10^{-6}$$

Plugging those numbers into the quadratic equation and doing the math:

$$x = \frac{-1.78 \times 10^{-5} \pm \sqrt{(-1.78 \times 10^{-5})^2 - (4 * 1 * - 1.78 \times 10^{-6})}}{2 * 1}$$

and

$$x = 1.34 \times 10^{-3}$$

Because of the +/− term for the root, there are two possible numerical solutions.
    The answers are:

$$8.9 \times 10^{-6} + 1.33 \times 10^{-3} \quad \text{and} \quad 8.9 \times 10^{-6} - 1.33 \times 10^{-3}.$$

Subtracting 1.33×10^{-3} would not make sense since that would give us a negative molarity, which is not practical. We discard that root and instead add 1.33×10^{-3}, giving an answer of:

$$1.339 \times 10^{-3} \text{ rounded to } 1.34 \times 10^{-3}.$$

As you can see, solving a quadratic equation takes some work. However, it is just basic algebra. In this case, with our concentration only listed as 0.1 molar and our K_a determined only to the hundreds place, variations in the thousands place is insignificant and imparts fairly unnoticeable error. (See Chapter 10 C for a review of significant figures.)

Weak Bases:

Obtaining the hydroxide concentration for a weak base is identical to solving the hydronium ion concentration for a weak acid. It simply requires solving the equilibrium expression for the K_b.

For our weak base ammonia:

$$K_b = \frac{[NH_4^+][OH^-]}{[NH_3]}$$

We can look up the K_b value for the base. We follow the same steps used for the weak acid problem, except that hydroxide ion is the unknown we seek. Hydroxide ion concentration, $[OH^-]$, is assigned the value x. The concentration of the conjugate acid $[NH_4^+]$ must be equal to the concentration of hydronium ion based upon the balanced equation:

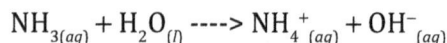

$$NH_{3(aq)} + H_2O_{(l)} ----> NH_4^+{}_{(aq)} + OH^-{}_{(aq)}$$

For each mole of hydroxide ion produced, one mole of cation is produced (this is true for 1:1 relationships). Therefore, $[NH_4^+]$ is also equal to x. Generally, we have the initial concentration of the intact acid NH_3. However, at equilibrium, some of that base has disassociated, so the equilibrium concentration is less than the initial concentration. How much less? x less. As a result, the concentration of NH_3 is *i-x*.

Substituting into our equation:

$$K_b = \frac{[x][x]}{[i-x]} = \frac{[x^2]}{[i-x]}$$

This is the same equation we obtained for our weak acid except for substitution of the K_b for the K_a.

Example 8-2:

What is the hydroxide ion concentration of a 0.2 M aqueous solution of ammonia?

The first step in solving this problem is obtaining the correct K_b value for ammonia. Ammonia has a K_b of 1.74×10^{-5}. The unknown is hydroxide ion concentration and its value is assigned x. From the balanced equation:

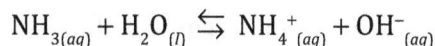

$$NH_{3(aq)} + H_2O_{(l)} \leftrightarrows NH_4^+{}_{(aq)} + OH^-{}_{(aq)}$$

There is a 1:1 molar relationship between the concentrations of hydroxide ion and ammonium ion (NH_4^+), so NH_4^+ is also equal to x. The initial concentration of ammonia is given as 0.2M, so at equilibrium, the concentration is reduced by x and can be represented by $0.2 - x$. Our expression becomes:

$$1.74 \times 10^{-5} = \frac{[x^2]}{[0.2 - x]}$$

This is a quadratic equation again, and we will make the same assumptions we made above in solving for our weak acid. Since the K_b is small (1.74×10^{-5}), this means that very little base has actually disassociated. Not much hydroxide ion has been produced, and we will make the assumption that the initial concentration of base is pretty much equal to the equilibrium concentration of base or:

$$0.2 - x \approx 0.2$$

With this substitution, our equation becomes:

$$1.74 \times 10^{-5} = \frac{[x^2]}{[0.2]}$$

and

$$x = 1.87 \times 10^{-3} \text{ M}$$

This is much simpler to solve. The K_b is multiplied by 0.2 and the square root is taken; the value of $x = 1.87 \times 10^{-3}$ M.

Now, you must always check the assumption you made to ensure it is true. You assumed that the concentration of ammonia at equilibrium was essentially the same as it was initially. Initially, it was 0.2 M. At equilibrium, it is $0.2 - x$, or $0.2 - 1.87 \times 10^{-3}$, which is 0.198 M. This is a difference of about 1 percent. So, making the assumption introduced about 1 percent error but saved some algebra. Again, if we needed greater accuracy, the quadratic equation could be solved as shown above for the weak acid example.

A note about temperature

Equilibria, and therefore equilibrium constants, are temperature-dependent quantities. Generally, K_a and K_b values are reported at STP. However, small changes in temperature can correspond to significant changes in the position of the equilibrium (a fancy way to say concentrations of products and reactants at equilibrium). This dependence can be calculated using the van't Hoff relationship. If you are looking for very accurate numbers, you must make sure that the K_{eq} you obtain is at the temperature at which you will conduct your work, and you must obtain the correct K_{eq}. You may find tabular equilibrium versus temperature data for a particular reaction, or you may need to apply the van 't Hoff relationship. If you are just looking for ballpark numbers, then small changes in temperature won't be problematic. In most laboratory work in the life sciences, the van 't Hoff relationship is rarely needed and isn't covered in this treatment.

SECTION C

Just What Are pH and Ka, and How about Kw?

BACKGROUND AND THEORY

For some reason, pH seems to confound students and laboratory technicians. This always surprises instructors because it is a simple concept; but students are often determined to make it complicated.

The concentration of acid in solution is commonly expressed in molarity. For solutions of strong acids, the hydronium ion concentration in aqueous solution at equilibrium is equal to the acid concentration. For weak acids, the concentration of hydronium ion is determined using the equilibrium expression and K_a (Part B of this chapter). However, these concentrations are generally small, even for strong acids. Using molarity to express the concentrations is correct; however, it is often inconvenient since numbers like 1.53×10^{-4} don't just roll off the tongue. To solve this, a scale was defined to provide more convenient numbers for these small molar concentrations. This scale is the pH scale, and it is simply the normal logarithm (base 10) of the molar concentration of H^+ or hydronium ion concentration. The log value is then multiplied by -1 to make the scale not only based on whole numbers, but also on positive numbers. The Danish chemist Soren Sorenson developed this scale in 1909. The abbreviation pH stands for *pondus hydrogenii*, Latin for "potential for hydrogen ion."

In math-speak:

$$pH = -\log[H_3O^+]$$

(Remember, brackets indicate M concentration.)

That's it! The pH scale is simply the molar concentration expressed logarithmically and multiplied by -1. (If you have forgotten all about logarithms, see Chapter 4.)

For our 1.53×10^{-4} M concentration, the pH is

$$pH = -\log 1.53 \times 10^{-4} = -1 \times -3.81 = 3.81$$

It is much easier to say or write (or definitely type) 3.81 than 1.53×10^{-4}.

The pH Scale

Because H^+ cannot exist freely in solution (it is tied to water) and instead exists as the hydronium ion (H_3O^+), there is a maximum and minimum concentration of hydronium ion that can be present in water. These concentrations are determined by the equilibrium constant of water. The minimum and maximum concentration of hydronium ion in water provides the practical upper and lower limits of the pH scale, and the negative logarithm of these molarities provides the common boundaries of the pH scale. The standard pH scale ranges from 0 to 14, corresponding to H^+ molarities from 1 to 1×10^{-14}, respectively. These boundaries are not absolute. Concentrations exceeding those commonly expressed on the pH scale do exist. However, the pH scale encompasses the vast majority of the hydronium ion concentrations present in the laboratory. Given this broad range of possible concentrations of H^+ in water, you can see how the pH scale of 1–14 seems much more compact and usable than molar concentrations.

The limits of acidity and basicity

The midpoint on the pH scale is 7. At pH 7.0, the concentrations of hydronium ion and hydroxide ion are equal (and equal to 1×10^{-7} M). We define the solution as neutral when the concentrations are equal. When the molar concentration of hydronium ion is greater than 1×10^{-7} M, the pH is lower than 7 and the acid is in excess compared to base. We call the solution acidic. Likewise, if the hydronium ion concentration is less than 1×10^{-7} M, the pH is greater than 7 and the solution is basic. (Remember that the effect of taking the negative logarithm of the concentration is that smaller numbers are higher on the scale.) We can understand this qualitatively, and we can also look at the situation quantitatively to determine from where these molar concentrations arise.

In water:

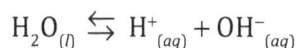

$$H_2O_{(l)} \leftrightarrows H^+_{(aq)} + OH^-_{(aq)}$$

More correctly:

$$2H_2O_{(l)} \leftrightarrows H_3O^+_{(aq)} + OH^-_{(aq)}$$

The equilibrium constant can be written for the forward reaction (the disassociation of water) as shown in Part B of this chapter:

$$K_{eq} = \frac{[H^+][OH^-]}{[H_2O]} = K_w$$

The equilibrium constant for water is given a special symbol K_w and has a value of 1×10^{-14} at STP. Since water is the solvent, its value is removed from the expression and the equation reduces to:

$$[H^+][OH^-] = K_w = 1 \times 10^{-14}$$

If the concentrations of hydronium and hydroxide are equal, solving the above equation for the molar concentration, x, provides a value of 1×10^{-7}.

$$x \times x = 1 \times 10^{-14}$$
$$x = 1 \times 10^{-7}$$

and

$$pH = -\log 1 \times 10^{-7} = 7$$

Likewise, in an acidic solution where the acid concentration, x, is ten times the base concentration, $0.1x$:

$$x \times 0.1\, x = 1 \times 10^{-14}$$
$$x = 3.16 \times 10^{-7}$$

and

$$pH = -\log 3.16 \times 10^{-7} = 6.5$$

Since the pH is less than 7, this is an acidic solution. We know this to be true since the acid is in tenfold excess compared to the base.

THE OTHER P VALUES

pOH

The pOH scale is parallel to the pH scale. It is also based upon the disassociation of water into hydronium and hydroxide ions. Like pH, the pOH is a log scale from 0–14 since it stems from the same equilibrium equation with the same limits. A low pOH value indicates a highly basic solution; pOH of 7 is neutral (equal concentrations of hydroxide and hydronium ions), and a pOH above 7 is not basic (and therefore acidic). As pOH increases, pH decreases, and vice versa. By far, pH is more commonly used than pOH.

pKa

Molar hydronium ion concentrations are inconvenient; the pH scale addresses this issue. Likewise, equilibrium constants, like K_a and K_w values, are also small and usually expressed using exponential notation. The mathematical operation of the negative logarithm is applied to hydronium ion concentration to provide a more convenient scale; it is also applied to the equilibrium constant in the same manner.

Therefore:

$$-\log [H^+] = pH$$

so:

$$-\log K_a = pK_a$$

and

$$-\log K_w = pK_w$$

pK_a values are often used rather than K_a values for simple convenience.

A trick of convenience–determining family relationships

A neat little trick is to use the rules of logarithms to take the negative log of the entire (simplified) K_w equation above:

$$-\log\{[H^+][OH^-]\} = -\log K_w = -\log 1 \times 10^{-14}$$

(at STP)
and

$$-\log[H^+] - \log[OH^-] = -\log K_w = -\log 1 \times 10^{-14}$$

simplified

$$pH + pOH = pK_w = 14$$

The negative log of the hydronium ion concentration (pH) plus the negative logarithm of the hydroxide ion concentration (pOH) in water is 14. This relationship is useful in conveniently obtaining the concentration of either the acid or basic species in water.

pK_b

This section has focused on the K_a and the pK_a; however, for bases, a similar treatment can be done simply substituting K_b and pK_b and hydroxide for hydronium ion. Further, the K_w expression shows us that in water, the K_w relates the pH and pOH. Additionally, in water, the pK_a and pK_b of an acid base pair (HA and A^-) are also related by the K_w:

$$pK_b + pK_a = pK_w$$

This is a convenient relationship if a desired equilibrium constant value is not available. The value can often be calculated by simple subtraction if its conjugate acid or base value is available.

In the Laboratory:

Most often, in the lab you will know either a molar concentration of acid and need to calculate the pH or vice versa. Examples of these calculations follow:

Calculating pH from the Molarity

Example 8-3:

Predict the pH of a 1.7×10^{-3} M solution of HCl in water.

Since HCl is a strong acid, the disassociation is assumed to be 100 percent complete, and the concentration of H_3O^+ is 1.7×10^{-3} M.

Using the definition of pH:

$$pH = -\log[H_3O^+] = -\log 1.7 \times 10^{-3} = 2.8$$

Reality Check: The solution is a strong acid in water and a fairly acidic pH should be expected; 2.8 is below 7, so therefore the answer makes sense.

And Back Again—Calculating Molarity from pH

To obtain hydronium ion concentration from pH, you use the same equation. The thing to remember here is that pesky little negative sign. You must not forget it!

$$pH = -\log[H_3O^+]$$

Rearranging the equation to solve for pH, you can see what happens to the negative.

$$-pH = \log[H_3O^+]$$

And solving for $[H_3O^+]$, you need to take the antilog (which in base 10 is simply 10^x)

$$10^{-pH} = [H_3O^+]$$

For our example above with a pH of 2.8:

$$10^{-2.8} = [H_3O^+] = 1.6 \times 10^{-3} \, M$$

A review of the rules of exponents and logarithms can be found in Chapter 4. Neglecting to change the sign of the pH is the most common mistake students make. However, if you perform a reality check, you will find your mistake.

Done incorrectly:

$$10^{2.8} = [H_3O^+] = 630 \, M$$

The answer is 630 M? That is a bit too concentrated and should let you know that you must have made a mistake.

Note: Comparison of the two examples shows a difference in concentrations. Why is the molar concentration slightly lower when calculated from a pH of 2.8? This is simple round-off error. The pH value was rounded up from 2.77 to 2.8. When the molarity was calculated from a pH of 2.8, a value of 1.58 was obtained because of the round-off. (For more information on correct rounding and error, see Chapter 10 C.)

For weak acids, obtaining the pH from a molar concentration of hydronium ion is identical to the problem for strong acids. However, if the concentration of the acid is provided instead of the hydronium ion concentration, the concentration of hydronium ion must be calculated using the equilibrium expression as shown in this chapter, Part B, weak acids. Then the pH can be determined by taking the negative logarithm of the hydronium concentration (x).

pOH

The concentration of hydroxide can be calculated in a parallel manner. For solutions of strong base, the concentration of hydroxide at equilibrium is equal to the initial concentration of base since the base is 100 percent disassociated. Then, pOH is simply the negative logarithm of the molar concentration of hydroxide ion.

$$pOH = -\log[OH^-]$$

Example 8-4:

Determine the pOH and the pH of a 0.1 mM solution of KOH.

Since the base is strong (a metal hydroxide), the concentration of hydroxide is equal to the initial concentration of base, KOH. But here, the concentration of base is expressed as mM. The relationship, which defines pH or pOH, holds only for molar concentrations. Therefore, the mM value must be converted to M (see Chapter 5 C for the definition of mM).

$$0.1 \text{ mM} \Rightarrow \frac{0.1 \text{ mmol}}{L} \times \frac{1 \text{ mol}}{1000 \text{ mmol}} = 0.0001 \frac{\text{mol}}{L} \Rightarrow 0.0001 \text{ M}$$

Substituting into the definition of pOH:

$$pOH = -\log 0.0001 = 4$$

In this case, the pOH is 4.

Reality Check: This is a fairly dilute solution of a strong base. Therefore, the pOH should be basic (less than 7), but not extremely basic because there isn't that much hydroxide ion around. The pOH of 4 seems reasonable.

Example 8-5:

What would be the pOH if you forgot to convert from mM to M in the problem in Example 8-4?

$$pOH = -\log 0.1 = 1$$

In this case, the pOH is 1.

This is a pretty basic solution for something so dilute.

To determine the pH from the pOH, it is useful to remember that K_a and K_b are related by K_w. Once disassociated in water, the hydroxide ion reacts and reaches equilibrium governed by the K_w and:

$$pH + pOH = pK_w = 14$$

Therefore:

$$pH = 14 - pOH$$

and the pH of this solution is:

$$14 - 4 = 10.$$

Reality Check: The pH of this solution of strong base is in fact basic (above 7).

For weak bases, obtaining the pH or pOH from a molar concentration of hydroxide ion is identical to the problem for strong bases, and the pH can be determined by subtraction from 14. However, if the concentration of the base is provided, the concentration of hydroxide ion must be calculated using the equilibrium expression as shown in Part B. Then, the pOH can be determined by taking the -log of the hydroxide ion concentration (x) and the pH determined by difference.

SECTION D

Monobasic versus Dibasic—What about Polyprotic Acids?

BACKGROUND AND THEORY

Titratable protons

As mentioned in the previous sections of this chapter, acids or bases can produce more than one hydronium ion or hydroxide ion in solution. Those acids that produce only one hydronium ion like HCl are termed monoprotic acids. However, many acids are di- or triprotic, and some may have even more titratable protons. The word *titratable* may bring back horrible memories of beakers, pH meters, and burettes filled with base in freshman chemistry lab. But pH curves are really quite simple to understand; even curves for triprotic acids like phosphoric acid (H_3PO_4) are simple when examined part by part. In fact, looking at a titration curve critically can help you really understand exactly what is going on in solution at each pH.

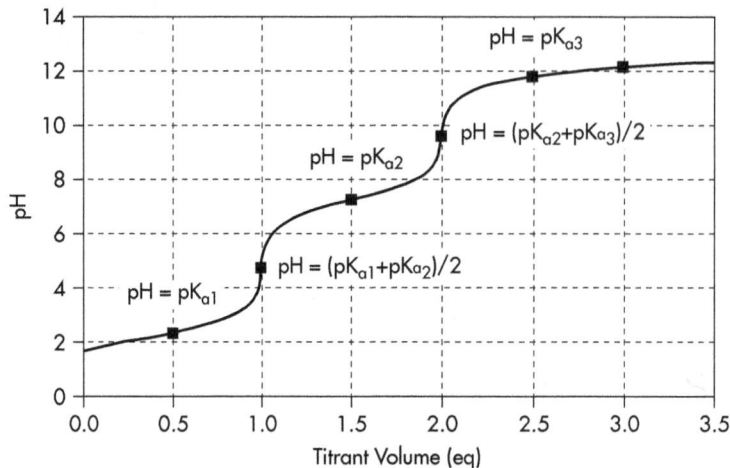

FIGURE 8-1 A typical titration curve for phosphoric acid.

In this graph, phosphoric acid has been added to a beaker (usually with water added as well) and titrated with a strong base like NaOH. Initially, the pH of the acid is low as would be expected for a strong acid in water undisturbed by the addition of any additional base. If the solution were simply the phosphoric acid in water, the disassociation of the acid would be governed solely by the K_a of the acid, and pH could be determined using the K_a values. The initial pH is below 2, which indicates an acidic solution, as would be expected.

Initially, phosphoric acid disassociates its first proton. The equilibrium is governed by the equation:

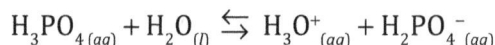

$$H_3PO_{4\,(aq)} + H_2O_{(l)} \leftrightarrows H_3O^+_{\,(aq)} + H_2PO_4^-{}_{\,(aq)}$$

This first disassociation is governed by the first K_a for phosphoric acid 7.5×10^{-3}. Phosphoric acid is not as strong an acid as HCl, but the K_a is substantial and the first proton disassociates, although not completely. The pK_a is 2.12, the negative logarithm of the K_a. The conjugate base of H_3PO_4 is $H_2PO_4^-$. This anion is commonly referred to as the monobasic form, since $H_2PO_4^-$ can add one equivalent of acid (the reverse reaction above).

As base is added to the aqueous acid during the titration, neutralization occurs:

$$H_3O^+_{\,(aq)} + OH^-_{\,(aq)} \leftrightarrows H_2O_{(l)}$$

The hydronium ion produced as the phosphoric acid disassociates is "consumed" by the added base and water is produced. The effect of the addition of base is to remove hydronium ion from the solution, and therefore the disassociation of the acid is pushed forward to produce more hydronium ion. The addition of base causes neutralization of liberated hydronium ion and promotes the disassociation of the acid, and the pH increases.

At the point where the pH equals the pK_a, it can be shown that the first proton is one half disassociated. (This doesn't mean that half a proton leaves; that isn't possible. It means that one half of all of the first protons are titrated.) This is the first midpoint of the titration. Inspection of the graph shows that it occurs in the middle of the first fairly horizontal portion of the graph at a pH slightly above 2 (2.2, the pK_a, to be exact).

Continued addition of hydroxide further increases the pH slowly as more hydronium ions are produced (governed by the first acid disassociation equilibrium constant, K_{a1}) and neutralized by the added hydroxide. This region of slow pH change near the pKa ($+/- 2$ units from the pK_a) is called the buffer region. In the buffer region, there are significant concentrations of both the conjugate acid and base; as a result, additions of acid or base cause relatively little change in pH.

The sharp break in the curve (vertical rise) occurs at the point where the first proton has been fully titrated (neutralized). This is the end of the titration of the first proton, and "all" H_3PO_4 is effectively gone and the monobasic form, $H_2PO_4^-$, predominates.

Continued addition of base begins to encourage the second disassociation governed by the equilibrium:

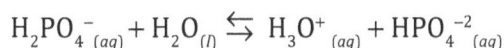

$$H_2PO_4^-{}_{(aq)} + H_2O_{(l)} \leftrightarrows H_3O^+{}_{(aq)} + HPO_4^{-2}{}_{(aq)}$$

This is the equilibrium where the monobasic phosphate $H_2PO_4^-$ disassociates to the dibasic phosphate HPO_4^{-2}. The conjugate base of the first disassociation becomes the acid of the second disassociation.

Further titration raises the pH to the midpoint of titration of the second proton, which occurs at the second pK_a, 7.2 ($K_a = 6.2 \times 10^{-8}$). This is the midpoint of the second somewhat horizontal portion of the curve and is noted on the graph. At this pH, the monobasic and dibasic forms of the acid are equal in concentration. This is also the midpoint of the second buffer region. The addition of more hydroxide finishes the titration of the second proton, and a steep rise in pH is seen.

The titration of the third proton occurs as even more hydroxide is added. It is governed by the equation:

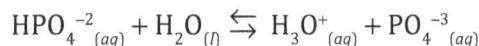

$$HPO_4^{-2}{}_{(aq)} + H_2O_{(l)} \leftrightarrows H_3O^+{}_{(aq)} + PO_4^{-3}{}_{(aq)}$$

This equilibrium has a very small K_a of 4.8×10^{-13}. The midpoint of the titration where the dibasic HPO_4^{-2} concentration equals the tribasic PO_4^{-3} concentration is reached at $pH = pKa_3$, or 12.3. This is the midpoint of the third horizontal portion of the graph and is also the midpoint of the third buffer region.

Further inspection of the graph (and a little added thought) should leave you thinking. Why isn't there a third steep rise corresponding to 100 percent disassociation of the third proton? The answer is that this requires a pH too high for water, and the third disassociation does not continue to completion.

In the Laboratory

If you carefully study the titration curve for phosphoric acid and keep in mind what is occurring at each point in the graph, you will gain a good understanding of acid base behavior in water. Few di- or polyprotic acids disassociate completely in water. To be able to determine hydronium ion or hydroxide ion concentration of a polyprotic acid or base, you must be able to determine which disassociation reaction is applicable at the pH of your solution. If you cannot do that, you will not be able to choose the correct pK_a for your solution preparation. This is extremely important when determining concentrations of buffers.

SECTION E

The Dreaded Buffer and the Infamous Henderson-Hasselbalch Equation

Many students do well with acid base calculations, but their brains come to a screeching halt when faced with a buffer problem. Hopefully, this section will destroy those mental barriers.

A buffer gets its name because it buffers or resists changes in its composition—namely, changes in pH. As you can see by looking at the titration curve for phosphoric acid, addition of base results in gradual increases in pH throughout most of the titration (except for at the titration endpoints). Phosphoric acid is a good buffer in three different pH ranges. Buffers resist changing pH when bases (or acids for that matter) are added. Buffers are extremely important biologically, as organisms, like you, are complex aqueous systems (held in a skin-and-bone container). It is critical that your pH be maintained despite the millions of reactions occurring in your body that both consume and produce hydronium ions.

A buffer is quite simple in composition. A buffer is composed of a weak acid and its conjugate base (often called its salt). If you understand what is happening in the titration of phosphoric acid with base, then you understand buffer behavior. In fact, the phosphoric acid buffer system is one of the most important biological buffer systems. Phosphoric acid has three distinct buffer regions. Each region centers around one of its pKa values: namely, 2.2, 7.2, and 12.3. At each equivalence point, the pH equals the pKa. At each pH, a different equilibrium reaction is occurring:

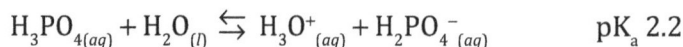

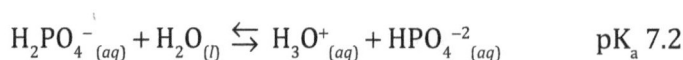

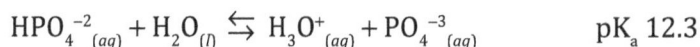

$$H_3PO_{4(aq)} + H_2O_{(l)} \leftrightarrows H_3O^+_{(aq)} + H_2PO_4^-_{(aq)} \qquad pK_a\ 2.2$$

$$H_2PO_4^-_{(aq)} + H_2O_{(l)} \leftrightarrows H_3O^+_{(aq)} + HPO_4^{-2}_{(aq)} \qquad pK_a\ 7.2$$

$$HPO_4^{-2}_{(aq)} + H_2O_{(l)} \leftrightarrows H_3O^+_{(aq)} + PO_4^{-3}_{(aq)} \qquad pK_a\ 12.3$$

Near each pK_a there are appreciable amounts of both the acid form and the salt form available. In fact, at the point where pH equals the pK_a (the equivalence point), there are equal concentrations of acid and conjugate base in solution. Therefore, near a pK_a, addition of base pushes the equilibrium to the right, but only slightly. Likewise, addition of acid pushes the equilibrium to the left, but only slightly. At a pH near the pK_a, there are enough acid and enough base to react with added acid or base to keep the equilibrium in nearly the same position. Additions of acid or base to a weak acid/salt solution result in small changes in pH. Essentially, you are "riding" the horizontal portion of the titration curve. If you add too much base or acid, you will go past the buffer region ("blow the buffer") and your buffer will no longer resist changes in pH. This is seen in the vertical portions of the titration curve where small additions of base (small changes on the x axis) result in large pH changes (y axis).

Choosing a Buffer

To create an aqueous buffer, you add a weak acid and its salt (its conjugate base) to water in defined concentrations to produce the pH desired. If the buffer is good, it will maintain that pH within its limits. A buffer should be chosen so that its pK_a is near the pH that needs to be maintained.

For the example above, phosphoric acid, there are three distinct buffering regions: 2.2, 7.2, and 12.3. It should be no surprise that the second pK_a—the one that is close to physiological pH—is the important equilibrium biologically. To make a phosphate buffer that buffers human physiological pH (7.34), the buffer should consist of $H_2PO_4^-$ and HPO_4^{-2}, the monobasic and dibasic forms, as active species. Most commonly, phosphate buffers are made from the potassium salts KH_2PO_4 and K_2HPO_4 or the sodium salts NaH_2PO_4 and Na_2HPO_4.

The monobasic potassium salt disassociates in water, producing the potassium ion and the monobasic anion.

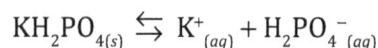

$$KH_2PO_{4(s)} \leftrightarrows K^+_{(aq)} + H_2PO_4^-{}_{(aq)}$$

The potassium ion is simply a spectator ion in solution and is ignored. While it contributes to the ionic strength of the solution, it does not appreciably affect the pH.

The monobasic anion reacts with water to produce hydronium ion, the acid disassociation reaction.

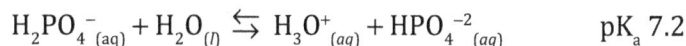

$$H_2PO_4^-{}_{(aq)} + H_2O_{(l)} \leftrightarrows H_3O^+_{(aq)} + HPO_4^{-2}{}_{(aq)} \qquad pK_a \ 7.2$$

The same is true for the dibasic salt.

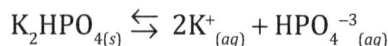

$$K_2HPO_{4(s)} \leftrightarrows 2K^+_{(aq)} + HPO_4^{-3}{}_{(aq)}$$

and

$$HPO_4^{-3}{}_{(aq)} + H_2O_{(l)} \leftrightarrows H_3O^+{}_{(aq)} + PO_4^{-3}{}_{(aq)} \qquad \text{pKa } 12.3$$

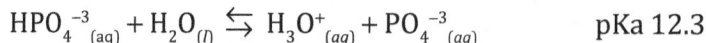

However, the high pK_a of the disassociation of the dibasic salt shows that near physiological pH, essentially none of the dibasic salt will react with water to produce hydronium ion. Thus, the predominant equilibrium occurring at physiological pH is the disassociation of the monobasic salt. This is the only equilibrium that must be considered at physiological pH.

In practicality, a buffer buffers extremely well within one pH unit of its pK_a. It continues to buffer, albeit more weakly, up to two pH units away from the pK_a. Remember, one pH unit is a concentration change of 10^1 or tenfold; therefore, two units is 10^2 or 100 times. A change in hydronium ion concentration equal to a two-unit pH change is a hundredfold change in hydronium ion concentration.

Determining the concentrations of the potassium salts needed to make a buffer is easily accomplished using the equilibrium constant for the weak acid. For buffers, the acid equilibrium expression has been algebraically manipulated to make it convenient to use. The result of this is the Henderson-Hasselbalch equation.

A derivation

Derivation of the Henderson-Hasselbalch Equation

While it isn't necessary to be able to derive the Henderson-Hasselbalch equation to use it effectively in the laboratory, it is instructive to study its derivation in order to understand it fully. Students are often very intimidated by the buffer equation; however, it is simply a weak acid equilibrium rearranged in base 10 log form. (If you don't remember the basic rules of logarithms, see Chapter 4 B.)

For a weak acid:

$$HA_{(aq)} + H_2O_{(l)} \leftrightarrows H_3O^+{}_{(aq)} + A^-{}_{(aq)}$$

and

$$K_a = \frac{[H_3O^+][A^-]}{[HA]}$$

To obtain the Henderson-Hasselbalch expression, the negative logarithm is taken of each side of the equation:

$$-\log(K_a) = -\log\left(\frac{[H_3O^+][A^-]}{[HA]}\right)$$

The negative logarithm of the K_a is simply the pK_a. The quotient on the right is expanded using the rules of logarithms.

$$pK_a = -(\log[H_3O^+] + \log[A^-] - \log[HA])$$

and

$$pK_a = -\log[H_3O^+] - \log[A^-] + \log[HA]$$

recognizing that $-\log[H_3O^+]$ is pH:

$$pK_a = pH - \log[A^-] + \log[HA]$$

Typically, the two log terms are combined. How you do this determines the form of the Henderson-Hasselbalch equation.

If the log term is added and the equation rearranged to solve for pH, the equation becomes:

$$pH = pK_a + \log\left(\frac{[A^-]}{[HA]}\right)$$

Or if the log term is subtracted:

$$pH = pK_a - \log\left(\frac{[HA]}{[A^-]}\right)$$

The Henderson-Hasselbalch equation allows you to determine the pH if the ratio of weak acid (HA) to conjugate base (A^-) is known. It also allows you to calculate the ratio of weak acid to salt that will maintain a specified pH.

This relationship is often stated more generally as:

$$pH = pK_a + \log\left(\frac{[base]}{[acid]}\right)$$

or

$$pH = pK_a - \log\left(\frac{[acid]}{[base]}\right)$$

Pay attention to terminology!

The latter statements of the Henderson-Hasselbalch equation often lead to problems of terminology. Base refers to the conjugate base or salt, A^-. The term *acid* refers to the intact (non-disassociated) weak acid, HA. Students often think of the *acid* as H^+ or as H_3O^+. But the hydronium ion term is included as the pH. So, substituting hydronium ion concentration for acid concentration leads to a mess.

Also, for a polyprotic acid like phosphoric acid, the conjugate base of the first acid disassociation (H_2A^-) becomes the acid for the second disassociation reaction. The acid H_2A^- then disassociates to the second conjugate base, HA^{-2}. Next, HA^{-2} becomes the acid for the third disassociation to form the final conjugate base, A^{-3}. By definition, an acid disassociates a proton to form a conjugate base. Whether a species is called an acid or a conjugate base depends upon which reaction you observe.

Limitations of the Henderson-Hasselbalch Equation

In using the Henderson-Hasselbalch equation, you make the assumption that only one equilibrium is occurring in solution, the reaction governed by the pK_a of the weak acid under study. You assume that the concentration of the acid and its conjugate base at equilibrium is that produced only by the acid disassociation (there are no other reactions "available" to any species in the reaction). What is ignored here is that the salt is a base and will be hydrolyzed in water. This will affect the position of the acid reaction to some extent. Also, by itself, water dissociates, providing some hydroxide and hydronium ions. At pH values near 7, these assumptions introduce little error, and the Henderson-Hasselbalch equation is a good tool for buffer calculations. However, if a stronger acid or stronger base is used, or if the solution is very dilute, the Henderson-Hasselbalch assumptions can lead to significant error. In the former case, the strength of the conjugate base cannot be ignored, and in the latter case, the disassociation of water becomes significant. Additionally, the Henderson-Hasselbalch relationship fails when used for a pH more than two units from its pK_a (where the buffer fails). This is because a greater than hundredfold concentration difference exists between acid or base in these regions, and hydrolysis of the salt becomes important. However, for most acids with pK_a's within a few units of neutrality and buffers in the mM to M range, the Henderson-Hasselbalch equation gives good results easily and quickly.

In the Laboratory

The section on background and theory in this chapter provided information on the reasons for the labeling of a particular buffer appropriate for a particular application. In general, a buffer is composed of a weak acid and its conjugate base. Both species must be highly soluble in water at the desired pH. In addition, the pK_a of the weak acid should be as close as possible to the pH that is to be maintained. A good buffer buffers very effectively within one pH unit of its pK_a. Some buffering will take place beyond one unit +/− the pK_a. In fact, you will see some good buffering (depending on concentration) up to +/− 2 units from the pK_a. Beyond that point, however, buffering is poor, and the solution will behave as either an acid solution or base solution.

For multiprotic acid buffers, you must examine the pK_a that is closest to the desired pH and choose the acid and salt accordingly. For our phosphoric acid example near physiological pH, the acid would be the monobasic salt, MH_2PO_4 (where M indicates a +1 metal like K^+ or Na^+), and the base would be the dibasic salt M_2HPO_4. Students are often tempted to use phosphoric acid H_3PO_4 as the acid. This would be technically appropriate only for a solution buffered near phosphoric acid's first equilibrium between H_3PO_4 and the monobasic salt, pKa 2.2. And given the limitations of the Henderson-Hasselbalch equation discussed above, using that equation will not give strictly accurate results at a pH so far removed from neutrality.

The best way to truly understand any equation is to apply it. Several example calculations using the Henderson-Hasselbalch equation follow.

Example 8-6:

Estimate the pH of a solution made by dissolving 8.2 g of monobasic sodium phosphate and an equal amount of sodium phosphate dibasic in 100 mL of water.

This is a buffer problem. The two salts are the active species for the second disassociation of phosphoric acid:

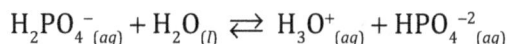

$$H_2PO_4^-{}_{(aq)} + H_2O_{(l)} \rightleftarrows H_3O^+{}_{(aq)} + HPO_4^{-2}{}_{(aq)}$$

The pKa for this reaction can be found and is generally reported as 7.2.

Since this is a reaction for a weak acid and its salt, it is a buffer problem. Using the Henderson-Hasselbalch equation will be a convenient way to solve it.

$$pH = pK_a + \log\left(\frac{[\text{base}]}{[\text{acid}]}\right)$$

In this instance, the pK_a is known, the pH is the desired quantity, and amounts of base and acid are given. Equilibrium expressions call for molar concentrations. So, for the base and acid, these concentrations must be calculated from the data given. See Chapter 5 for examples of the calculation molarity if you are unsure how to proceed.

For NaH_2PO_4, 8.2 g are dissolved in 100 mL. The formula mass of the monobasic salt can be calculated as 119.98 g/mol. NaH_2PO_4 is the acid in this reaction. The molarity is:

$$M = 8.2\text{ g} \times \frac{1\text{ mol}}{119.98\text{ g}} \times \frac{1}{100\text{ mL}} \times \frac{1000\text{ mL}}{L} = 0.68\text{ M}$$

For Na_2HPO_4, the data are the same, except that the formula mass is 141.96 g/mol. Na_2HPO_4 is the base in this reaction:

$$M = 8.2 \text{ g} \times \frac{1 \text{ mol}}{141.96 \text{ g}} \times \frac{1}{100 \text{ mL}} \times \frac{1000 \text{ mL}}{L} = 0.58 \text{ M}$$

Using the Henderson-Hasselbalch equation:

$$pH = pK_a + \log\left(\frac{[\text{base}]}{[\text{acid}]}\right)$$

Note: By the balanced equation, the monobasic salt serves as the acid and the dibasic salt as the base in this equilibrium. Substituting in the values:

$$7.2 + \log\left(\frac{0.58}{0.68}\right) = pH$$

and

$$pH = 7.13$$

Reality Check: This problem illustrates a very important property of buffers, one that will help you check your buffers to make sure your answers are realistic. When the pH equals the pK_a, the concentration of acid and conjugate base are equal. The buffer is at the "height" of its buffering capability with strong concentrations of reactant acid and product base. Here, the molar concentrations of the acid and conjugate base are nearly equal. The acid is slightly higher than the base (0.1M greater); therefore, the pH should be lower than—but nearly equal to—the pK_a. It will save you a lot of time, error, and probably money in the lab if you know what to expect from your calculation before you do the calculation. Reality checks will help you make a solution correctly the first time!

Note: The most important part of any buffer calculation is correctly identifying the base, the acid, and the correct pK_a. It can help to write the chemical equation. That should be the first step in any buffer calculation. Once written, you will be able to easily identify the acid and the base from the balanced reaction. If you can do that, then the rest is just using the equation.

Example 8-7:

Five hundred mL of a 0.1 M potassium phosphate buffer at pH 7.4 is needed.

This is a buffer problem dealing with the second disassociation of phosphoric acid. When calculating how to make a buffer in the lab, this example is the most practical to follow.

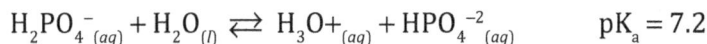

$$H_2PO_4^-{}_{(aq)} + H_2O_{(l)} \rightleftarrows H_3O+{}_{(aq)} + HPO_4^{-2}{}_{(aq)} \qquad pK_a = 7.2$$

Here, the desired pH is given. What is needed are the concentrations of acid and salt. There are two variables in the Henderson-Hasselbalch equation: the molar concentration of the base and of the acid.

$$pH = pK_a + \log\left(\frac{[base]}{[acid]}\right)$$

One equation and two variables is a problem. However, there is a relationship provided that solves that conundrum. The desired buffer concentration is 0.1M. That means that the concentration of acid and base must total 0.1M.

$$[base] + [acid] = 0.1M$$

And this relationship can be solved for either acid or base; here, it is solved for [base].

$$[base] = 0.1M - [acid]$$

And this expression substituted into the Henderson-Hasselbalch equation provides a single equation with a single variable.

$$pH = pK_a + \log\left(\frac{0.1 - [acid]}{[acid]}\right)$$

Substituting numbers into the expression:

$$7.4 = 7.2 + \log\left(\frac{0.1 - [acid]}{[acid]}\right)$$

Subtracting and taking the antilog:

$$1.6 = \frac{0.1 - [\text{acid}]}{[\text{acid}]}$$

Solving for [acid]:

$$1.6[\text{acid}] = 0.1 - [\text{acid}]$$
$$2.6[\text{acid}] = 0.1$$
$$[\text{acid}] = 0.1/2.6$$
$$[\text{acid}] = 0.039\text{M}$$

using the expression:

$$[\text{base}] = 0.1\text{M} - [\text{acid}]$$

the concentration of the base can be determined:

$$[\text{base}] = 0.1\text{M} - 0.039\text{M}$$
$$[\text{base}] = 0.061\text{M}$$

Reality Check: The desired pH is above the pK_a. Therefore, more base is present than acid. The concentration of the base in this calculation is higher than the acid concentration. Thus, the numbers are on track.

In this example, you have determined the molar concentrations of the monobasic (acid) and dibasic (base) forms of potassium phosphate that when mixed together will give you the desired pH. However, you can't stop here; you need to know how to make the solution. This is no longer a buffer problem—it is now a molarity problem.

An important tip

To proceed, you need to know something about the chemicals you will be using. Both the monobasic and dibasic forms of potassium phosphate are available as white solids. For this case, I have assumed the solids are anhydrous and used the formula masses for the anhydrous solids. If hydrates are available in your laboratory, you will need to use the formula mass for the correct solid (for hydrates, see Chapter 3 C). To make a molar solution from a solid, you need the volume and the formula mass to determine the mass needed.

For the acid, potassium phosphate monobasic, KH_2PO_4, the concentration determined above is 0.039 M; the desired solution volume is 500 mL; and the calculated formula mass is 136.09 g/mol:

$$0.039 \text{ M} \Rightarrow \frac{0.039 \text{ mol}}{\text{L}} \times \frac{136.09 \text{ g}}{1 \text{ mol}} \times \frac{1 \text{ L}}{1000 \text{ mL}} \times 500 \text{ mL} = 2.65 \text{ g}$$

For the base, potassium phosphate dibasic, K_2HPO_4, the desired concentration is 0.089 M with the same solution volume; the formula mass is 174.18 g/mol:

$$0.061 \text{ M} \Rightarrow \frac{0.061 \text{ mol}}{\text{L}} \times \frac{174.18 \text{ g}}{1 \text{ mol}} \times \frac{1 \text{ L}}{1000 \text{ mL}} \times 500 \text{ mL} = 5.31 \text{ g}$$

To make the buffer, weigh 2.65 g of KH_2PO_4 and 5.31 g of K_2HPO_4 into a flask and dilute to nearly 500 mL with distilled water. Check the pH and adjust if necessary. Add water to obtain 500 mL.

It is generally necessary to adjust the pH of your buffer

You have just made your buffer. You are beaming with pride. You measure the pH, and it's not correct! Don't be heartbroken. Some adjustment is usually necessary. The pK_a varies with temperature and in this problem was obtained to only one decimal place. There is rounding error in determining the molarities and the masses. Plus, you may weigh 2.62 g of acid rather than 2.65 g, etc. Remember the limitations of the Henderson-Hasselbalch equation in that only a single equilibrium is considered. That imparts some error to your calculation. The pH you obtain will most likely differ somewhat from the desired 7.4. This is no problem. Simply adjust the pH with acid (most appropriately, phosphoric acid) or base (potassium hydroxide, most appropriately) until the desired pH is obtained. In fact, you could make your buffer by starting with phosphoric acid and titrating up to the desired pH with KOH. But this would be extremely inefficient!

Example 8-8:

One liter of a 0.5 M sodium acetate buffer at pH 5.2 is to be prepared.

This is an acetic acid buffer. Acetic acid has only one proton to disassociate, so choosing the pKa is easy:

$$CH_3CH_2COOH_{(aq)} + H_2O_{(l)} \rightleftarrows H_3O^+_{(aq)} + CH_3CH_2COO^-_{(aq)}$$

Acetic acid is often abbreviated HOAC and the acetate ion OAc⁻.

The pK_a for this reaction can be looked up and is generally found to be about 4.75. The desired pH is given. Therefore, as in the previous example, what is needed are the concentrations of acid and salt. Using the Henderson-Hasselbalch equation,

$$pH = pKa + \log\left(\frac{[\text{base}]}{[\text{acid}]}\right)$$

The buffer concentration desired is 0.5M. This means that the concentration of acid and base must total 0.5M.

$$[\text{base}] + [\text{acid}] = 0.5\text{M}$$

and

$$[\text{base}] = 0.5\text{M} - [\text{acid}]$$

Substituting into the Henderson-Hasselbalch equation:

$$pH = pKa + \log\left(\frac{0.5 - [\text{acid}]}{[\text{acid}]}\right)$$

Substituting numbers into the expression:

$$5.2 = 4.75 + \log\left(\frac{0.5 - [\text{acid}]}{[\text{acid}]}\right)$$

Subtracting and taking the antilog (10^x) of both sides:

$$2.82 = \frac{0.5 - [\text{acid}]}{[\text{acid}]}$$

Solving for [acid]:

$$2.82[\text{acid}] = 0.5 - [\text{acid}]$$
$$3.82[\text{acid}] = 0.5$$
$$[\text{acid}] = 0.131 \text{ M}$$

and solving for [base]:

$$[base] = 0.5 \text{ M} - [acid]$$
$$[base] = 0.369 \text{ M}$$

Reality Check: The desired pH is above the pKa. Therefore, there should be a greater concentration of base than of acid. Here, the concentration of the base is higher than the acid, so the numbers are on track.

You have determined the molar concentrations of the acetic acid and the sodium acetate that will give you the desired pH. You next need to know how to make the solution.

For acetic acid, the calculated concentration is 0.131 M and the desired solution volume is 1 L; the formula mass is 60.0 g/mol. To go farther, you must know the physical forms of your acid and base because you will need to measure them. Acetic acid is not available as a solid. If you simply plug and chug without thinking, you will use the formula mass to calculate an amount to weigh. But even though you can weigh a liquid, you will not be weighing a pure liquid if you weigh concentrated acetic acid. Acetic acid is available as an aqueous solution. Hence, if you weigh the solution, your mass is the sum of the acetic acid and water. You will therefore not be adding the amount of acetic acid you need.

Various concentrations of acetic acid are available, and you would do best to use the one you find in your lab. Most commonly, acetic acid is sold in its concentrated form or in an anhydrous form called glacial acetic acid. Glacial acetic acid is almost 100 percent water free. For this example, assume you have a bottle of concentrated acetic acid in the lab. If you look up the concentration of concentrated acetic acid, you will find it is 17.4 M. So, you need to make a 0.131 M solution from a 17.4 M solution. This is a dilution problem (Chapter 7). Using the dilution equation and solving for the desired volume, V_1:

$$C_1 V_1 = C_2 V_2$$
$$17.4 \text{ M } (V_1) = 0.131 \text{ M}(1\text{L})$$
$$V_1 = 0.00753 \text{ L}$$
$$\text{or } 7.53 \text{ mL}$$

For the base, sodium acetate, the calculated concentration is 0.369 M and the desired volume 1 L; the formula mass is 82 g/mol. Sodium acetate is a solid.

$$0.369 \text{ M} \Rightarrow \frac{0.369 \text{ mol}}{\text{L}} \times \frac{82 \text{ g}}{1 \text{ mol}} \times 1 \text{ L} = 30.3 \text{ g}$$

To make the buffer, weigh 30.3 g of NaOAc into a one-liter flask and dissolve by the addition of sufficient distilled water. Slowly add 7.53 mL of concentrated acetic acid. Check the pH and adjust if necessary. Add water to obtain a final volume of 1 L.

Example 8-9:

In the laboratory, 250 mL of a 0.3 M potassium citrate buffer at pH 7.0 is needed.

Again, this is a buffer problem dealing with a polyprotic acid. Citric acid is a triprotic acid, formula $C(OH)(CH_2COOH)_3$ with pK_a values of 3.13, 4.76, and 6.40.

The desired pH is given. The third or last acid disassociation with a pK_a of 6.40 is the pKa closest to the desired pH of 7.0. The concentrations of the correct acid and salt are needed to make the buffer. In this case, the dibasic salt is the acid and the tribasic salt is the base.

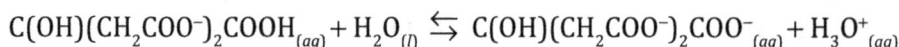

$$C(OH)(CH_2COO^-)_2COOH_{(aq)} + H_2O_{(l)} \rightleftharpoons C(OH)(CH_2COO^-)_2COO^-_{(aq)} + H_3O^+_{(aq)}$$

Using the Henderson-Hasselbalch equation:

$$pH = pK_a + \log\left(\frac{[base]}{[acid]}\right)$$

and the relationship for buffer strength:

$$[base] + [acid] = 0.3 \text{ M}$$

And this relationship can be solved for either acid or base; here, it is solved for [base].

$$[base] = 0.3 \text{ M} - [acid]$$

and this expression substituted into the Henderson-Hasselbalch equation.

$$7.0 = 6.40 + \log\left(\frac{0.3 - [acid]}{[acid]}\right)$$

Subtracting and taking the antilog (10^x) of both sides:

$$3.98 = \frac{0.3 - [acid]}{[acid]}$$

Solving for [acid]:

$$3.98[acid] = 0.3 - [acid]$$
$$4.98[acid] = 0.3$$
$$[acid] = 0.060 \text{ M}$$

and:

$$[base] = 0.3M - 0.06 \text{ M}$$
$$[base] = 0.24 \text{ M}$$

Reality Check: The desired pH is slightly above the pK_a. Therefore, there should be a greater amount of base present than of acid. The concentration here of the base is higher than the acid, so we know the numbers are on track.

You have determined the molar concentrations of the dibasic (acid) and tribasic (base) forms of potassium citrate that will give you the desired pH. However, you can't stop here; you need to know how to make the solution. This is no longer a buffer problem; it is now a molarity problem. Examining the shelf (or a catalog, the Internet, or a handbook) shows that both the dibasic and tribasic salts are available as solids. Anhydrous solids are assumed for this problem.

For the acid, potassium citrate dibasic, $K_2C_6H_6O_7$, the calculated concentration is 0.06 M. The desired solution volume is 250 mL; the formula mass is 268.30 g/mol:

$$0.060 \text{ M} \Rightarrow \frac{0.060 \text{ mol}}{L} \times \frac{268.30 \text{ g}}{1 \text{ mol}} \times \frac{1 \text{ L}}{1000 \text{ mL}} \times 250 \text{ mL} = 4.02 \text{ g}$$

For the base, potassium citrate tribasic, $K_3C_6H_5O_7$, the desired concentration is 0.24 M with the same solution volume; the formula mass is 306.42 g/mol:

$$0.24 \text{ M} \Rightarrow \frac{0.24 \text{ mol}}{L} \times \frac{306.42 \text{ g}}{1 \text{ mol}} \times \frac{1 \text{ L}}{1000 \text{ mL}} \times 250 \text{ mL} = 18.38 \text{ g}$$

To make the buffer, weigh 4.02 g of $K_2HC_6H_5O_7$ and 18.38 g of $K_3C_6H_5O_7$ into a container and dilute to dissolve in distilled water. Check the pH and adjust if necessary. Add distilled water to a final volume of 250 mL.

A shortcut? It depends ...

The above examples show how to make a buffer by calculating the ratio of acid to base and adding them to obtain a desired pH (with a little adjustment). In the laboratory, it is often the practice to simply add the salt in the molarity desired for the buffer and titrate with acid to the correct pH. This will produce a buffer with the correct pH. However, it may be more time consuming, and the buffer concentration will not be correct. But depending on your application, it might not matter.

Also, for many buffers, buffer recipes can be found on the Web or in the literature. Using these saves some calculation time, although if you practice with buffers, the calculation is quite rapid. Literature recipes can most likely be used with confidence. However, it is always best to check recipes found on the Internet. They aren't always correct, and if you don't know how to calculate a buffer, you won't be able to determine if the recipe is your problem when a procedure goes wrong.

CHAPTER 9
Multicomponent Assays

SECTION A

It All Must Add Up

BACKGROUND AND THEORY

Often in the laboratory, your task is to perform an assay that contains several chemical species; each "test tube" or sample contains several solutions, each at a specified concentration. Tasks of this type are often referred to as multicomponent assays.

Success is achieved one variable at a time

There are several general principles to remember when performing multicomponent assays, or assays that contain more than one chemical species. The first rule of thumb is that in order to determine the effect of a single condition—for instance, changing concentration of a substrate or enzyme in a kinetics reaction or changing concentration of magnesium ion in a PCR reaction—you must vary only one solution concentration at a time. All other concentrations (other than the one being studied) and all other conditions (time, temperature, etc.) must be identical. Otherwise, you cannot be certain what caused the observed change.

Solutions dilute each other

Secondly, you have to realize that when you add more than one solution together, the solutions dilute each other. The final concentration of a particular component in the assay tube is not identical to the concentration you initially added to the assay tube. The final volume of each sample is additive (the sum of the volumes of each component). Since volume is the denominator in molar concentrations, increasing the volume decreases the concentration. If you want a certain final concentration of a particular species, you must add that species in a more concentrated form in order to ensure that the final concentration is correct. You must account for the dilution upon mixing with other reagents. The needed stock concentration can be easily calculated using the dilution equation (Chapter 7). This chapter will deal with the SI unit molarity; however, the same principles can be applied to other concentration units.

Multicomponent assays are really quite easy (although occasionally time consuming) to set up and should not present a significant challenge if you approach the setup systematically. If you can make molar solutions, use the dilution equation, and make a buffer (for buffered assays), you have every skill needed to accomplish the task. This section will provide you with an organized approach to setting up assays and keeping track of all component concentrations.

SECTION B

A Simple Table Is the Way to Go

BACKGROUND AND THEORY

The best way to approach setting up an assay is to make a table that contains a column for each chemical species in the assay and a row for each sample needed. You should then add an additional column for water or buffer and another column to keep track of final volume. You can then fill in the table and use that table as the protocol for assembling the assay samples. You can even add a column and check off the boxes in your laboratory notebook as you add each sample. This can keep you from forgetting a component. If you perform the task in a spreadsheet, then you can use cut/paste and fill functions to automate the task and save a lot of calculator time.

	A	B	C	D	water/buffer	Total
Sample 1						
2						
3						
4						
5						
6						

For example, for an assay that contains four chemical species (A through D) and 6 samples, the table would resemble the one above:

Of course, your table would contain units! The hardest part of setting up an assay is the beginning. You need to decide your final volume (total volume) of each sample. This is often prescribed. In the example above, the final volume would be entered into the final column. The total volume must be the same for each sample.

	A	B	C	D	water/buffer	Total
Sample 1						V_t
2						V_t
3						V_t
4						V_t
5						V_t
6						V_t

Next, you must decide what final concentrations of each reagent are needed (this is frequently prescribed as well). You must make or buy the stock of each component, A through D, at a concentration sufficient that once diluted in each sample, the final concentration will be the desired concentration. In a way, you are starting at the end (final concentration in the assay tube, C_2) and working backward to the initial or stock concentration (C_1).

You must keep in mind that the volumes of each component A through D and the volume of water or buffer added must sum to the total volume for the assay tube. You need to choose a volume of each component to add to each sample that is measurable and since you are choosing the volume to add, you might as well make it convenient! As with serial dilutions, choosing multiples of 10 makes it easier to keep track of dilutions. For example, if your total assay tube volume was 1 mL (1000 µL), you could choose to add 100 µL of a component. This would give a dilution factor of 1:10 in each sample and indicate that the stock would need to be tenfold more concentrated.

The concentration of one component varies while all other concentrations remain constant. Changing concentration for that component can be most easily accomplished by simply adding different volumes of the same stock to each assay tube. To maintain the same total concentration (V_t), the volume of water or buffer (see the note below about choosing water or buffer) must also change.

In this case, component D was chosen to vary. As the volume of the varied component increases, the volume of water or buffer decreases. The volume of water or buffer needed can be obtained by subtraction.

Our table becomes:

	A	**B**	**C**	**D**	**water/buffer**	**Total**
Sample 1	V_A	V_B	V_C	V_1	$V_t - (V_1 + V_C + V_B + V_A)$	V_t
2	V_A	V_B	V_C	V_2	$V_t - (V_2 + V_C + V_B + V_A)$	V_t
3	V_A	V_B	V_C	V_3	$V_t - (V_3 + V_C + V_B + V_A)$	V_t
4	V_A	V_B	V_C	V_4	$V_t - (V_4 + V_C + V_B + V_A)$	V_t
5	V_A	V_B	V_C	V_5	$V_t - (V_5 + V_C + V_B + V_A)$	V_t
6	V_A	V_B	V_C	V_6	$V_t - (V_6 + V_C + V_B + V_A)$	V_t

Once you have your table set up, you will need to determine the concentrations of each stock solution (A through D) to make. This is done using the dilution equation. Each reagent added dilutes to the other reagents to a total volume of V_t. Therefore, if you determine the dilution factor, it can be used to determine the concentration of the needed stock solution. The dilution factor (see Part C of Chapter 7) is simply the volume of a component divided by the final or total volume. For example, for component A, the dilution factor (df) would be V_A/V_t. To determine the stock solution concentration, the final concentration should be divided by the dilution factor.

This sounds confusing, but by using a number example you will be able to see that it is quite simple.

Example:

If V_A was chosen as 100 µL and V_t is 1 mL (1000 µL), the dilution factor is:

$$df = 100\ \mu L/1000\ \mu L = 0.1$$

This means that the concentration of component A is diluted tenfold (multiplied by 0.1) in each sample tube by the addition of the other reagents. Thus, the stock of component A would need to be made ten times more concentrated than the desired final concentration to account for the dilution of V_a to V_t.

So,

$$V_{A,stock} = V_{A,final}/df$$

Using our numbers from the example above:

$$V_{A,stock} = V_{A,final}/0.1$$

Or

$$10 \times V_{A,stock} = V_{A,final}$$

The same logic is applied to each reagent. The process for stock A need be repeated for stocks B and C.

Use only one stock, if at all possible!

Only one concentration need be calculated for each component—in other words, only one stock should be made for each component. Students are often tempted to add the same volume of the varying reagent (V_D in this case) but accomplish the different concentrations by making different stocks. This is a very poor approach for two reasons. First, it requires making six stocks instead of one stock, thus using more reagents and more time. Second, it introduces sources of error. In making six solutions, there is more possibility of technician or student error, and each sample comes from a different stock, introducing uncertainty about the consistency of composition of each sample.

If the volume you choose to add to each assay tube produces a stock that is too dilute to make, then adjust the volume of stock you add in each assay tube to be able to change the dilution factor and make a more convenient stock. As long as you can effectively and accurately measure the volume, its choice is up to you. Remember the upper limit of total assay volume and that you need to add the other reagents. If your stock is still too dilute to make, then you can either make more of that stock or use the serial dilution approach (Chapter 7).

Once you have determined the desired concentration for each component stock solution, then you must calculate how to make each stock. Generally, this is accomplished using the standard methods for preparing molar solutions (Chapter 5).

Buffer or water?

In a biological assay, the buffer concentration also needs to be constant (unless the buffer concentration is the variable under study). Consequently, if buffer is added as an independent component of the assay, then water—not buffer—must be added in order to maintain a consistent concentration of buffer in each sample. However, another choice is to make each stock solution in buffer (at the desired final concentration of buffer) rather than in water. In this case, the buffer, and not water, should be added to maintain consistent buffer concentration. The choice is often yours. If your samples are stable in water, you can make them in water and add a constant volume of stock buffer as a reagent. Then, you would dilute to the final sample volume with water. If your samples are best made in buffer, then make all samples in buffer and dilute to the final volume with buffer for each sample. However you choose, water or buffer, you must be consistent to avoid problems with varying concentrations of buffer.

In the Laboratory

A practical example of a kinetics assay setup used in my introductory biochemistry laboratory follows.

Example 9-1:

A student needs to collect data to determine the Michaelis-Menten constant (if you don't know what this is, no big deal; what we are doing will work for the setup of any multi-component assay) and maximum velocity for the human insulin tyrosine kinase under physiological conditions. Six data points should be collected. The protocol calls for the following final conditions and concentrations of reactants:

Reaction volume, 3 mL; citrate buffer, 100 mM; pH = 6.9; tyrosine substrate varied from 10 mM–to–100 mM; ATP 30 mM; enzyme–10 μg/mL.

In a Michaelis-Menten kinetics experiment, the concentration of the substrate (in this case, tyrosine) should be varied while all other concentrations and conditions are kept constant.

Following the method outlined above, a table should be created as follows:

	ATP	Enzyme	Buffer	Tyrosine	Water	Total Volume
Sample 1						
2						
3						
4						
5						
6						

The total volume of each sample is 3 mL, so that can be added to the table; for consistency, the volume was expressed as 3000 µL:

	ATP	Enzyme	Buffer	Tyrosine	Water	Total Volume (µL)
Sample 1						3000
2						3000
3						3000
4						3000
5						3000
6						3000

Here, the choice was made to prepare the stock reagents in distilled water. Therefore, buffer is chosen as an independent component of the assay. Convenient volumes of 300 µL to give a 1:10 dilution (300 µL /3000 µL) of each constant reagent were chosen.

The table changes as follows:

	ATP (µL)	Enzyme (µL)	Buffer (µL)	Tyrosine	Water	Total Volume (µL)
Sample 1	300	300	300			3000
2	300	300	300			3000
3	300	300	300			3000
4	300	300	300			3000
5	300	300	300			3000
6	300	300	300			3000

Again, the reagents are made up in distilled water. Thus, buffer must be added as a component of the assay.

The tyrosine concentration should vary from 10 to 100 mM as prescribed in the protocol. For convenience, sample 1 volume was chosen to be 100 µL and sample 6 to be 1 mL (1000 µL). Samples 2 through 5 have volumes varied to provide varied final concentrations. The exact volumes chosen were arbitrary.

The table becomes:

	ATP (μL)	Enzyme (μL)	Buffer (μL)	Tyrosine (μL)	Water	Total Volume (μL)
Sample 1	300	300	300	100		3000
2	300	300	300	200		3000
3	300	300	300	400		3000
4	300	300	300	600		3000
5	300	300	300	800		3000
6	300	300	300	1000		3000

The volume of water to add is determined by subtraction of all reagent volumes from the total volume. Water is used to maintain the final volume.

Our final table is:

	ATP (μL)	Enzyme (μL)	Buffer (μL)	Tyrosine (μL)	Water (μL)	Total Volume (μL)
Sample 1	300	300	300	100	2000	3000
2	300	300	300	200	1900	3000
3	300	300	300	400	1700	3000
4	300	300	300	600	1500	3000
5	300	300	300	800	1300	3000
6	300	300	300	1000	1100	3000

This table can be used in the laboratory to assemble the samples efficiently.

You aren't quite done yet!

To complete the preparation, the concentration of each stock must be determined. Since 300 μL was chosen as the volume for ATP, enzyme and buffer and the total volume of each sample is the same, the dilution factor is the same for each component:

$$df = V_i/V_f$$
$$= 300\,\mu L/3000\,\mu L = 0.1$$

So, each stock must be made at ten (1/0.1) times the desired final concentration.

For tyrosine, our varied substrate, the smallest volume added (sample 1) yields the lowest desired concentration. The dilution factor for sample 1:

$$df = V_i/V_f$$
$$= 100\ \mu L/3000\ \mu L = 0.033$$

Using the highest volume of stock for the most concentrated tyrosine sample, sample number 6:

$$df = V_i/V_f$$
$$= 1000\ \mu L/3000\ \mu L = 0.33$$

The dilution factor for the tyrosine varies tenfold from 0.033 to 0.33.

Making another table can help keep track

For each sample, the stock concentration needed is the final concentration for each sample divided by the dilution factor.

Component	Final desired assay concentration	Stock concentration needed (final conc./df)
ATP	30 mM	30 mM/0.1 = 300 mM
enzyme	10 µg/mL	10 µg/mL/0.1 = 100 µg/mL
buffer	100 mM	100 mM/0.1 = 1000 mM = 1 M
tyrosine*	10 mM	10 mM/0.033 = 303 mM

*The smallest concentration of tyrosine is chosen to make the stock. Increasing tyrosine concentration in each assay is accomplished by adding more (greater volumes) of the stock, not more concentrated stock.

How much should you make?

You next must determine how much of each reagent you need to make. This can be done for each reagent by summing the column in the assay table. You will typically want to make more than the minimum to allow for repeating a sample, poor pipetting, or spillage. You may need to make more to produce a manageable volume. Pay attention to waste and cost in deciding your volumes. Also, if you are going to perform the experiment in duplicate or triplicate, you will need to remember to multiply your final volumes appropriately.

Adding an additional row to our assay table:

	ATP (μL)	Enzyme (μL)	Buffer (μL)	Tyrosine (μL)	Water (μL)	Total Volume (μL)
Sample 1	300	300	300	100	2000	3000
2	300	300	300	200	1900	3000
3	300	300	300	400	1700	3000
4	300	300	300	600	1500	3000
5	300	300	300	800	1300	3000
6	300	300	300	1000	1100	3000
Total	1800	1800	1800	3100		

At this point, you need to determine how to make each stock shown in the stock table from the materials available in your particular laboratory. The physical form of each compound (solid, liquid, or solution) will determine how to proceed for each stock solution. The appropriate methods for making the solutions can be found in Chapters 5 and 6.

CHAPTER 10
Measurements

SECTION A

When Measurements Matter—the Really Practical Side

BACKGROUND AND THEORY

Don't waste your time!

Just how accurate do you need to be? That is a question that is seldom asked by students or new technicians in the laboratory. It is a very important question! Knowing just how accurate your measurements must be will dictate how you make those measurements. One approach I commonly see students take is to be as accurate as possible at all times. However, this is a foolish approach because accuracy takes time and careful handling. If you spend five times the required time to obtain a measurement that is five times more accurate than you need, you have wasted four times the time.

Generally, most common "wet methods"—those measurements made with burettes, volumetric pipettes, and volumetric flasks—are accurate to about a part per hundred or thousand. You have other, less accurate, means of measuring volume like graduated cylinders. The limitations of common laboratory glassware are listed in Table 10-1. Additionally, electronic balances can measure routinely to three decimal places; however, some measure to four and even five places. Electronic balances can be found in many laboratories.

In the Laboratory

In the laboratory, I observe two very common and diametrically opposed student approaches to accuracy. Occasionally, the same student making the same solution uses both approaches! The first is to very carefully measure the mass of a needed solid to the fourth decimal place exactly—meaning, if a calculated mass is 1.0007 g, the student painstakingly measures exactly 1.0007 g. This requires much fiddling at the balance: shake some on, take a tiny bit off, shake on, etc., until the exact mass is reached. This can take an inordinate amount of time. Then I have observed the same student add that excruciatingly measured solid to a beaker and add distilled water to the volume mark on the beaker.

What's wrong with that? The volume notations on a beaker are broad estimates and are not guaranteed to any level of accuracy (about +/− 1 ml). The student took an inordinate amount of time to very carefully and accurately measure a prescribed mass to the hundred thousandths place and then made the solution accurate to probably the tenths place at best when the volume was measured. Essentially, all of that time at the balance was wasted, and in the end, the solution was not made very accurately.

The most important first question ...

How do you avoid this trap? The first thing you must do in the laboratory is to ask the question "How accurate do I need to be?" You need to be aware of the parameters of your measurement. The next thing you need is an understanding of what methods and equipment will provide your needed accuracy (Part B, this chapter). Finally, you must pay attention to the data obtained and interpret it and express it correctly (Part C in this chapter).

Common sense is best

Choosing the correct measurement method often involves inspection of the experimental parameters and simply using some common sense. Actually, common sense is your best tool. At times, accuracy is specified (that makes your task an easy one). If a desired concentration is provided with a +/− suffix, you will know how accurate you need to be. For example, if a concentration is specified as 0.1M +/− 0.01, then you will have to make sure that all measurements are certain to the hundredths place. Additionally, a concentration could be expressed as 0.100 M. The addition of the 0 in the thousandths place implies that the measurement should be accurate to that decimal place (more on this in Part B, significant figures).

But what if you are making a buffer and the concentration is specified simply as 0.1 M? It is probably perfectly fine to make a buffer that is 0.10 M, or one that is 0.09 M, or one that is 0.011 M. Keep in mind that, unfortunately, protocols often require more accuracy than stated. People generally don't follow the rules (Section B) and instead make assumptions. For example, stating 0.1 M implies that a range as large as 0.06 M to 0.14 M will suffice for the experiment. Both of these values would correctly round off to the stated goal of 0.1 M. However, this broad range of concentrations may be acceptable—or it may not.

If faced with a protocol that asks for a 0.1 M buffer (100 mM), I will generally make a solution keeping the uncertainty in the hundredths place. In other words, a buffer that varies from 90 mM to 110 mM. This is easily done using an electronic balance and volumetric glassware. There is no reason to try and make an "exactly" 0.1000 M buffer. Making a buffer to this level of accuracy would require precise measurement of solids to the fourth decimal place and all volumes to the same (which would mean obtaining accurate densities and weighing the liquids in a typical laboratory). This would take significant time, time that is wasted, where the specification was only to the tenths place.

SECTION B

What Should I Use to Measure the Volume? Graduated Cylinders, Glass Pipettes, Pipettors, and Volumetric Flasks

Determining what instrument to use to measure volume depends on the accuracy needed. Table 10-1 below shows data for the most common volumetric glassware. Usually, if glassware is graduated, then it is assumed that the user can accurately estimate to one-tenth of the smallest graduation. Some users do this well; some do not. It is generally a matter of practice.

All about pipettes

Nominal Values versus True Values

The volume listed on a piece of glassware is called its nominal value. The true value is the actual quantity measured. Often, the true value is slightly different from the nominal value. For example, a 10 mL nominal value volumetric flask may actually be calibrated to deliver 10.04 mL. So, is the nominal value correct? Well, that depends on how correct (how accurate) you need to be. If you need to deliver 10.0 mL, then the true value of 10.04 mL is just as correct as 10.00 for your use. If you need very high accuracy, you will have to calibrate your glassware to know the true value.

TABLE 10-1 Limitations of Some Common Glassware

Instrument	Graduation Increments	Measurement Accuracy	Note
beaker Erlenmeyer flask	10 mL	+/− 1 mL	Graduations vary, depending on size of cylinder; best for rough estimation only
graduated cylinder	1 mL	+/− 0.1 mL	Graduations vary, depending on size of cylinder
burette	0.1 mL	+/− 0.01 mL	Graduations vary, depending on size of burette
volumetric flask	Not graduated	Accuracy varies based upon volume and class. Generally between +/− 0.1 mL and +/− 0.01 mL	
volumetric pipette (fixed volume)	Accuracy varies based upon volume and class. Usually between +/− 0.01 and +/− 0.001mL. Fixed-volume pipettes are generally more accurate. Fixed-volume pipettes incur less user error because final decimal place does not require user estimation.		
volumetric pipette (graduated)			
pipettor	1 µL to 1 mL depends on volume of pipettor and manufacturer. See manufacturer specifications for +/− accuracy values.		
transfer and Pasteur pipettes	Generally not graduated	Not calibrated for any level of accuracy	Don't use to measure volumes

What about the last drop of liquid in the pipette?

TC versus TD

When using a volumetric pipette, it is important to know whether the pipette is *TC* ("to contain") or *TD* ("to deliver"). For a TC pipette, all liquid must be ejected from the pipette to obtain the stated volume. In other words, the pipette contains the entire volume and the pipette must be fully emptied or "blown out."

TD pipettes retain some liquid in the tip when emptied. TD pipettes should never be blown out, but instead allowed to simply drain. TD pipettes are calibrated to retain a specific volume in the tip while delivering the required volume. It should be noted, however, that TD pipettes are generally calibrated with water at 20°C. If a different temperature is present in the laboratory, error (albeit small) will be introduced. Also, if a different liquid other than water is used, error will also be introduced. The error introduced will depend on how far the surface tension of the liquid deviates from the surface tension of water. In either case, TC or TD, liquid should be drained as appropriate from the pipette with the tip of the pipette touching the side of the receiving container. Thus, any drop hanging at the tip that is included in the measurement will be transferred.

Calibrating Glassware

Since mass can be measured definitively and accurately to up to five decimal places (four is more common), calibration is best accomplished using an analytical balance. Also, the accuracy of the analytical balance can be measured against a series of carefully measured weights that will not change mass or composition over time, ensuring consistent calibration of the balance. The calibration of a balance can be done by a lab worker or often by a balance technician.

Glassware is generally calibrated using water. The mass of a delivered volume of water is recorded, and water's density is used to calculate the volume actually delivered. That value can be compared to the nominal value. Of course, the density must be known to a degree of accuracy appropriate for the calibration. Since density varies with temperature, the temperature must also be measured and taken into account to obtain an accurate value.

Technically, two corrections must be made to the calibration for maximum accuracy. The first is a correction for the mass of an object weighed in air versus the mass when weighed in a vacuum. The second is the thermal expansion of the container itself. However, these corrections are very small and only needed in the most accurate analytical measurements; they are generally neglected.

The Correct Use of Pipettors

One of the biggest sources of error in the biochemical and biological laboratory is introduced by improper use of pipettors. Pipettors are convenient for delivering small volumes and are available as single channel (one pipettor) or multichannel (several pipettors "banded together"). They can be manual or automatic.

Read the instructions

When using a new pipettor, take a minute to actually read the calibration and use information provided by the manufacturer. For single-channel manual pipettors, the following rules of thumb address common user errors. They will help you avoid many errors in accuracy and precision:

1. Warm up the pipettor: depress and release the plunger several times before use.
2. Premoisten the tip several times with the volume of liquid to be delivered.
3. Wipe the tip only if there is liquid on the outside of the tip. Do not wick liquid from the inside of the pipette tip.
4. Don't blow bubbles! When aspirating (drawing up liquid), depress the plunger to the first stop before putting the tip in the liquid to be drawn.
5. When dispensing (letting out liquid), depress the plunger to the first stop and then past the stop to release the remaining liquid.
6. Use a smooth technique. Don't pop up the plunger. Depress and release in a slow, smooth motion.

7. Put the pipette tip straight in and pull straight out. Angles introduce error.
8. Immerse the tip below the meniscus of the liquid and don't touch the walls of the container (sides or bottom).
9. Always use the correct tip for the correct pipettor. You can't mix and match and obtain accurate volumes.
10. Don't lay a wet pipettor down on the bench. Liquids can flow back into the mechanism and damage the pipettor. Keep pipettors stored vertically!

SECTION C

Sig Figs—Your Calculator Lies

BACKGROUND AND THEORY

One problem with calculators is that they lie. By that, I don't mean that they provide an incorrect numerical answer based upon the input numbers and operations. I mean that they often provide a degree of accuracy that doesn't exist.

For example, let's say you measure the mass of 2.0 mL of a liquid and obtain 2.07 g mass from your balance. You can use your calculator to find the density by dividing the mass by the volume. Your calculator will provide the answer as 1.035 g/mL (the calculator will not provide the units; I added those). The problem with this answer is that it implies a degree of accuracy that simply doesn't exist. You measured your mass to the hundredths place and your volume to the tenths place. But your answer is somehow accurate to the thousandths place according to your calculator. This simply isn't possible. Common sense should tell you that no value could be more accurate than the least accurate measurement that was a part of that value. You can't create accuracy from inaccuracy. Unfortunately, students often don't use common sense and instead take the 1.035 g/mL answer as correct.

When common sense fails, rely on the rules

If your common sense fails you when it comes to determining the correct degree of accuracy, there is a system of rules that governs how measurements should be expressed. The rules determine the number of significant figures or significant digits allowed for any data obtained by calculation. Sig figs (significant figures) mystify freshman chemistry students. However, if you keep in mind the goal of sig figs, to make sure that a number reflects the accuracy of measurements used to create it, sig figs are really quite simple. Rather than memorizing a seemingly endless list of rules, you can learn significant figures based upon common-sense reasoning. Sig figs are simply a set of rules to keep your calculator from getting out of control and implying a detail in measurement that doesn't exist!

Before beginning a discussion of significant figures, it is important that you understand what accuracy (and its buddy, precision) really are. **Accuracy** reflects how close a measured value is to the correct or true value. **Precision** shows the agreement of measurements with each other. You can have a single measurement that is accurate. But in that case, you have no ability to measure precision. A measurement can be accurate and precise, the best-case scenario, or precise, but not accurate. The latter case usually implies a reproducible problem with the method used. You may be making the same error consistently. However, if a measurement is neither accurate nor precise, it is time to look for another method to accomplish the task.

The system of significant figures really deals with precision more than accuracy. The hope is that if data are reproducible or precision is good, then the data are accurate. This is true unless you have a reproducible error. If you have poor precision, your data can only be accurate by pure luck! If a measurement is not reproducible, it isn't useful. To determine accuracy, you always need a second method to validate the results of the first.

The zero rules:

Leading zeros are not significant.
Internal zeros are significant.
Trailing zeros are significant to the right of the decimal point.
Trailing zeros to the left of the decimal point are not significant (unless they are internal zeros).
When expressed in scientific notation, all zeros are significant.

Mathematical Operations:

For mathematical operations, it is important to understand the difference between position and number. By **position** means by the location of the digit. For instance, 1.1 contains a digit in the tenths place. Thus, this number has a significant digit in the tenths place, or tenths position. By **number** means the total number of significant figures. Our 1.1 contains two significant figures.

Addition or Subtraction:

The number of significant digits is determined by the measurement with the least number of significant figures by position.

Multiplication or Division:

The number of significant digits is determined by the measurement with the least number of significant figures by count.

In the Laboratory

The Eight Commandments of Significant Figures:
It is often easiest to understand the rules by example.

1. Leading Zeros Are Not Significant

Three significant figures are in 0.125: the 1, 2, and the 5 by number, and they are significant to the thousandths place by position. The leading 0 (to the left of the decimal) is simply a placeholder and is not significant. This makes common sense because this zero is written (correctly, I might add) simply to make sure you notice the decimal point.

2. Internal Zeros Are Significant

The number 0.1035 has four significant figures: the 1, 0, 3, and the 5 to the right of the decimal place by number. They are significant to the ten thousandths place by position. The internal 0 is significant because it is between two significant digits. Here, the internal zero is part of the measurement. The leading zero is still a placeholder and is not significant.

3. Trailing Zeros Are Significant to the Right of the Decimal Point

There are four significant figures in 0.1250: the 1, 2, 5, and 0 to the right of the decimal point by number. By position, they are significant to the ten thousandths place. The trailing zero is significant because by being listed, it implies that the measurement was actually made to the ten thousandths place. The value just happens to be 0. The leading zero is still a placeholder and is not significant.

4. Trailing Zeros to the Left of the Decimal Point Are Not Significant (Unless They Are Internal Zeros)

There is only one significant figure by number in 1,000: the 1. By position, it is significant only in the thousands place. However, 1001 has four significant figures by number, and by position is significant in the ones place. The first number, 1,000, implies that only the thousands place was accurately measured.

5. When Expressed in Scientific Notation All Zeros Are Significant

If your measurement of 1,000, above, was made to the ones place, and you want to make sure that the zeros are counted, it is best to use scientific notation and express your measurement as 1.000×10^3. If the measurement was accurate only to the thousands place, the correct expression in scientific notation would be 1×10^3. Scientific notation shows only significant zeros. In scientific notation, all zeros are significant; insignificant zeros "disappear."

6. Addition or Subtraction—by Position

The number of significant digits is determined by the measurement with the least number of significant figures by position. Using our numbers from above—0.125, 0.1035, and 0.1250—our rule tells us that when we add or subtract these numbers, that position rules. The least position in these numbers is the thousands position. Therefore, the answer can have significant figures only to the thousands position (the least accurate measurement). This is easily seen if you line up the numbers like you did in elementary school.

$$\frac{\begin{array}{r} 0.125 \\ + \ 0.1035 \\ 0.1250 \end{array}}{0.353\cancel{5}} = 0.354$$

There is a digit "missing" from the first datum, 0.125; therefore, the answer must be rounded to least significant or last fully filled position.

7. Multiplication or Division—by Number
The number of significant digits is determined by the measurement with the least number of significant figures by number.

For 0.125×0.1250, the calculator provides an answer of 0.015625. However, the answer can contain only three significant figures because 0.125 has the least significant figures, 3, by number. As a result, 0.015625 is best expressed as 0.0156, rounded off to three significant figures.

If you are questioning the significance of the 0 to the right of the decimal place, express the answer in scientific notation: 0.015625 becomes 1.5625×10^{-2}. It is rounded off correctly to three digits, or 1.56×10^{-2}.

What do you do with 5?

8. Rounding
It is generally better to keep an extra digit in multistep calculations and round off at the end rather than at each step in the calculation. This minimizes round-off error. Of course, numbers greater than 5 are rounded up, and those less than 5 are rounded down. But what do you do with the 5? In elementary school, you might have been taught to round 5 up. But there is a problem with this. If you always round up, you will skew your data consistently higher. So, how do you round 5? There are several schools of thought or conventions; two are presented here. The first is that if the digit before the 5 is even, round up; and if it is odd, round down, or vice versa. Just be consistent in your rounding. The reason for this is that over a set of calculations, you will most likely split rounding up and rounding down fairly evenly, minimizing round-off error. The second approach is to round up this time and round down next time. That sounds simpler. But the question is, do you remember which way you went—up or down—last time?

In the Laboratory

It is important to record your data accurately. If you measure to the fourth decimal place, you should record to the fourth decimal place. If you are using a four-decimal place balance and measure exactly one-tenth of a gram, it should be recorded as 0.1000 g. The leading zero is a placeholder so you don't miss the tiny little decimal point and read the data as 1000 g. The three trailing zeros indicate that the measurement was recorded to the fourth decimal place. Both are important—the leading zero for clarity and trailing zeros for accuracy.

It is just as important to pay attention to significant figures in calculations. When calculating using measured data, you may carry an extra digit and round at the end of the calculation. But don't record every digit that your calculator window displays. Do not let your calculator increase the accuracy of your measurements!

CHAPTER 11
What Not to Do with Biological Molecules
The Pitfalls of "Rough" Handling

BACKGROUND AND THEORY

Chemical handling and storage in the laboratory is, unfortunately, something rarely explicitly taught in laboratory courses. Students are provided with handling and disposal information as part of laboratory protocols; however, students are rarely taught a common-sense approach to general handling of chemicals. This is often better handled in an industry where there is an intense focus on safety. But I have seen industry technicians practice poor handling techniques with biological and biochemical samples because of basic ignorance about the samples they are working with.

Most importantly, for general safety, every student or worker should know how to access and read MSDS (Material Safety Data Sheets) materials in their laboratory. These sheets are required of each chemical manufacturer and cover the safe handling and disposal of the chemicals they sell. By law, MSDS materials must be available; luckily, today MSDS are available readily on the Internet.

There are some general guidelines, however, for dealing with biological or biochemical samples. First, you must know the hazards of any biological sample. If you are working with a virus or bacteria that is hazardous to human health, you should follow all prescribed precautions associated with the appropriate BSL, or Bio Safety Level. BSL or biocontainment levels are generally listed numerically from one to four. Level one substances are generally nonhazardous to human health and can be handled safely in a laboratory with only a chemical fume hood. Gloves, and possibly a simple filter mask, may also be prescribed for certain bacteria or viruses. General good laboratory practices should be used, but exposure to level one substances is generally nonhazardous. The BSL levels increase through level four. Level four substances must be fully contained and isolated. This means building systems, sealed rooms, sealed containers, and isolation of equipment must be practiced. Level four containment requires technicians to use containment suits and air supplies when working in these "hot zones." Any individual working at level four must undergo intense special training, including certification.

In this section, I will deal only with essentially nonhazardous or level one and below substances. Hazard levels greater than one require specialized training beyond the scope of this work. Most of what a life sciences student will encounter in his or her graduate or undergraduate career will involve level one or below hazards. Most often with level one or lower samples, there is a greater concern of you contaminating the sample rather than the sample contaminating you.

In the Laboratory

Chemical Hoods versus Biological Hoods

Surprisingly, many students do not know the difference between a chemical hood and a biological hood. Chemical hoods, or laminar flow hoods, are designed to draw air from the room through the air intakes and vent out the top into an air-handling system. The worker is protected from fumes when the hood is properly used because the user is "upwind" of the sample. These are the types of hoods you use in general chemistry with acids and bases and with volatile organics in organic chemistry laboratories. These hoods are designed to protect the user from the chemical. When using a chemical fume hood, you should keep the chemical under use within the confines of the hood and at least several inches to a foot (exact distance depends on the particular size of the hood and the rate of airflow) in front of the sash (door). Placing the sample on the lip of the hood or right inside the sash reduces the effectiveness of the hood, as the airflow is diminished at the face. You should open the sash the minimum distance to allow comfortable and safe work. Keep the sash closed far enough to protect your face at all times.

A biological hood is designed to contain the sample and to protect the sample from contamination by you and your environment. The direction of airflow in a biological hood is different from a chemical fume hood. In a biological hood, air is taken in through the air-handling system and exhausted through the hood and out toward the sash and therefore toward the user. Thus, your "cooties" will not infect the sample since you are downwind from the sample. Biohoods are not all filtered the same. The type of handling depends on the biohazard level certified for use in the hood. However, all biohoods are designed to keep and minimize contamination from the room and the user. This is not the intent of a chemical fume hood.

Handling Biochemical Compounds

Easy does it!

Many biochemical compounds can be treated the same as traditional chemicals. However, most proteins, nucleic acids, and some lipids require special care. A unique aspect of biochemical samples is that not only is the primary structure important for proper function, but also the higher-order structure is essential. Rough handling of biochemical samples will often not affect primary structure (break covalent bonds). However, higher-order structure is less rigid than that maintained by covalent bond forces and can be lost (and thus function lost or compromised) with poor technique. Without the higher-order structure, your sample may in fact be useless. To maintain efficacy, biomolecules must be handled with an eye to maintaining proper temperature and reducing other environmental stresses.

It is not difficult to determine how to handle a particular biological sample. Generally, it involves reading the bottle label. If a particular protein label indicates that the sample should be kept at 4°C, then the sample should be kept on ice while being used and returned to the freezer as soon as possible. Leaving the sample by the balance for hours or on top of the freezer will most likely destroy, or at least significantly alter, the sample's structure and thus its effectiveness. It is not necessary, however, to keep every sample in the biochemical laboratory on ice. In fact, many are stable at room temperature. It is worth a moment's thought and investigation about how to handle/store a sample before using it. Taking unnecessary precautions is time consuming and could actually cause sample problems; freezing, for example, may damage some samples.

Care should be taken when thawing and refreezing biomolecules. While many proteins, nucleic acids, and lipids are quite robust and can be completely thawed and refrozen indefinitely, some suffer severe damage by repeated or rapid freeze-thaw cycles. Usually, gradual changes in temperature are preferable to rapid changes and take less of a toll on higher-order structure. When removing an aliquot from a frozen sample, care should be taken to return the bulk of the sample to the freezer before it thaws.

Many biological molecules are subject to degradation by sheer stress. In other words, you can stir them to death! While it is perfectly acceptable to vigorously stir a salt solution or a buffer, you should generally avoid that with biological samples. Many samples can be safely and quickly vortexed, but some require gentler handling and should experience no more agitation than that provided by a rocker. While you want to ensure adequate mixing, many proteins can be denatured by too much sheer stress, and lipids can form emulsions that are difficult to resolve. When in doubt, be gentle with samples of biomolecules.

Rapid changes in pH should also be avoided. When changing pH of a biomolecular sample, the change should be done incrementally, giving the sample a chance to equilibrate between acid or base additions. Where possible, more dilute acids and bases rather than their concentrated cousins should be used (of course, acid or base concentration may be dictated by the dilution produced). Rapid decreases or increases in pH by bolus (all at once) injections of concentrated acid or base may cause destruction of higher-order structure and should be avoided wherever possible.

The rule of thumb for biomolecules is to treat them gently. In addition, it is critical that you pay attention to the manufacturer's instructions on correct storage and treatment. Failure to do so will often render your sample useless.

CHAPTER 12
Some Basic Statistics

BACKGROUND AND THEORY

When can I (legitimately) ignore those data?

Confidence Intervals, Limits, and Levels

Before applying statistical treatment of a data set, the desired confidence limit or confidence level should be determined. Confidence intervals, as opposed to point estimates, show a range of values for the mean of a given population or set of data. The interval shows the range that contains the true value within, providing a lower and upper limit for the data. The confidence interval is qualified by the confidence level, typically expressed as a percentage like 90 percent or 95 percent confidence. But what exactly does a 95 percent confidence level mean? It means that if a particular measurement is replicated, the value obtained would fall within the confidence interval (between lower and upper confidence limits) 95 percent of the time.

Simple Percent Error

The percent error is a useful comparison of a sample set mean to a known or accepted value. The calculation can be done quickly and provide a good approximation of accuracy. However, to calculate a percent error, the actual or accepted value must be known.

$$\% \text{ Error} = \frac{\text{Experimental value} - \text{theoretical value}}{\text{theoretical value}} \times 100$$

Standard Deviation

Standard deviations, also called root square variance, should be calculated for data sets wherever appropriate. It measures the data "spread." A standard deviation is a quick method to make a general decision about the validity of a data set. For example, if the standard deviation for a replicate measurement approaches the mean value, the data are probably not worth much. Depending on the required precision, the value of the standard deviation provides an assessment of the data's reproducibility. Standard deviations are easily calculated using spreadsheet programs or by many scientific calculators. However, standard deviations are fairly easy to calculate manually if those resources are not available. While the calculation can be tedious, especially with a large data set, it is not difficult. The formula for standard deviation (often symbolized by the Greek character σ) measures the deviation of each datum x_i from the mean $\bar{x}$. These differences are squared and summed. The square root of the summation is taken and the value divided by the number of data points, n, sans 1.

$$s = \sqrt{\frac{\sum_{i=1}^{n}(x_i - \bar{x})}{n - 1}}$$

Once calculated, the standard deviation should be appended to the data set mean. Data should be reported as value +/− standard deviation value. This process is illustrated for a very small data set in Table 12-1. Note that standard deviations measure variance or precision. A good standard deviation does not guarantee accuracy, just precision.

When can I ignore "bad" data?

One question that arises quite often in the analysis of data is, "Can I ignore a 'bad' data point?" The answer is, it depends. There are two legitimate ways to ignore data.

Always keep a record of your observations

The first legitimate method to discard a data point is because of a justified data collection problem. If an error or potential error is experienced in the collection of data, it should be noted in the laboratory notebook. For example, in setting up an assay, you might be unsure if enzyme was added to a particular test tube, or maybe you weren't sure that all of a particular reagent had been expelled from a pipette because the volume of that tube looks different from the others. If you choose to continue the experiment with the questionable sample, a note should be made (and dated) in the laboratory notebook. Later, if data collected from the sample are anomalous, then the data can be legitimately discarded because of the noted error. It is not appropriate, however, to Monday-morning quarterback and discard data without justification. Sure, you "might" have forgotten to add enzyme and that might explain the bad data. But if you recognized a potential mistake at the time and didn't record that in your notebook when the sample was made, then you cannot speculate and guess at what might have happened and discard data at will.

The second method to use when you need to discard data is to use statistical methods to determine whether the particular datum is in fact an outlier. There are numerous statistical methods that can be used to determine the legitimacy of data. Some treat all data with equal weight and justify discarding a bad data point (Q test). Others assign differing weights to data points rather than discarding a point—for example, Grubbs' test. Details of more complicated statistical methods are beyond the scope of this work. Students or technicians are encouraged to study further if more complex methods are needed to treat data.

Using the Standard Deviation to Examine and Discard Data

It is generally accepted that a data point that is more than 2.7 standard deviations from the mean may be discarded. This value has less than a 1 percent chance of being caused by inherent experimental fluctuations and is more likely due to a mistake in data collection. Thus, the data point can be discarded and the mean and standard deviation recalculated without the anomalous datum. No more than one data point should be discarded for any given data set. If more than one data point is discordant, then the data set should be questioned and possibly additional measurements made.

Grubbs' Test for Outliers

Like the example above, using the standard deviation to examine data, Grubbs' test searches the data for an outlier. Grubbs' test is also known as the maximum normalized residual test (that's why we use Grubbs'!). Only one outlier per data set may be examined. The absolute value of the datum with the maximum difference from the mean is divided by the standard deviation of the data set. The test statistic generated, the g value, is compared to a critical g value that depends on the level of confidence desired (typically 99 percent or 95 percent) and the number of data points in the data set. If the value of g calculated is less than the g_{crit}, then the data cannot be discarded. Grubbs' tables are readily available on the Internet and in statistics handbooks.

$$g = \frac{\max\limits_{i\,=\,1\ldots n} \left| x_i - \bar{x}_i \right|}{s}$$

The Q Test

The Q test, like Grubbs' test, is also used for identification and rejection of outlying data points. The Q test should be applied a maximum of once for a particular data set. In determining a data point's Q value, the data must be arranged in order of increasing values, and Q is calculated as absolute value of the quotient of the gap and the range of data.

$$Q = |gap/range|$$

The gap is the difference between the value in question and its closest numerical value. The range is the difference between the lowest and highest data points obtained in the data set. The quotient is obtained and its absolute value taken to provide the Q value. The Q value is then compared to a Q table, and if $Q_{data} > Q_{critical}$, the datum can be legitimately discarded. $Q_{critical}$ is dependent upon the number of values in the sample. Tables of critical Q values can be found in statistics books or on the Internet. Tabular Q values are provided at varying confidence levels. The choice of an appropriate confidence level depends on the particular application.

The Coefficient of Determination, R²

The R^2 value can be calculated for values that are plotted to produce a line of best fit. It is a representation of how well the data fit the line. The coefficient of determination represents the percent of the data that is closest to the line of best fit. In spreadsheets, this value is calculated automatically when a trend line is applied to a plot. An R^2 value of 1 indicates all values fit the line and there is no variation. An $R^2 = 0.95$ indicates that 5 percent of the variation cannot be explained by the line. The further the points diverge from the line, the lower the R^2 and the lower the confidence in the data.

In the Laboratory

No data set should go unchallenged. I am continually amazed that students or technicians often fail to even look at the data obtained and examine it for legitimacy. In the academic laboratory, students collect data, wait one or two weeks, and the night before the laboratory report is due, they attempt to analyze the data and write the report. It is often at this point that the student realizes that "the data aren't any good." You should never leave the laboratory without at least a quick review of the collected data to determine if they seem valid. In other words, are the trends in the direction you expected? Is the standard deviation reasonable if you have replicate measurements? If the data show a linear or higher-order trend, is the R^2 (square of the correlation coefficient) reasonable for the particular analysis? If you cannot answer these questions, you are unprepared.

Example 12-1:

Calculate the standard deviation for a data set composed of eight masses: 2.00 g, 2.10 g, 1.82 g, 2.00 g, 1.92 g, 2.05 g, 1.98 g, and 1.96 g.

TABLE 12-1 Sample Standard Deviation for a Small, n = 8, Data Set. Given the Calculated Standard Deviation, the Mean 1.98 Should Correctly Be Expressed as 1.98 ± 0.08.

x	$\bar{x}$	$x - \bar{x}$	$(x - \bar{x})^2$	$\sum(x - \bar{x})^2$
2.00	1.98	0.02	0.0004	0.0497
2.10		0.12	0.0144	$\sum(x - \bar{x})^2 / n - 1$
1.82		−0.16	0.0256	0.0071
2.00		0.02	0.0004	$\sqrt{\sum(x - \bar{x})^2 / n - 1}$
1.92		−0.06	0.0036	0.0842
2.05		0.07	0.0049	Number
1.98		0.00	0	1.98 ± 0.08
1.96		−0.02	0.0004	

Example 12-2:

Using the standard deviation, determine if there is an outlier in Table 12-1.

Examination of the data shows the most likely potential outlier in the data set is 1.82. This datum looks suspect. Applying the rule above ($2.7s$) to our standard deviation:

from Table 12-1, our standard deviation is 0.08,

$$2.7 \times 0.08 = 0.216$$

$$\bar{x} = 1.98; \ \bar{x} - 0.216 = 1.76; \ \bar{x} + 0.097 = 2.19$$

Our suspect datum is 1.82, and 1.82 is between the interval of 1.76 to 2.19; the point may not be discarded.

Example 12-3:

Apply the Grubbs' test to the data set in Table 12-1.

Examination of the data shows that the maximum deviation from the mean is caused by the point 1.82.

The deviation is −0.16, and the absolute value is 0.16. Our standard deviation, s, is 0.08.

$$g = 0.16/0.08 = 2.00$$

Examination of Grubbs' tables for n = 8 (our number of data points) shows that the g_{crit} for $\alpha = 0.05$ is 2.1266 and for $\alpha = 0.01$ is 2.2744.[1] Our value of 2.00 is less than the g_{crit} for both the 99 percent and 95 percent confidence levels. Therefore, we are stuck with our 1.82, and it remains in the data set.

Example 12-4:

Apply the Q test to the data set in Table 12-1

To apply the Q test, the data should be arranged in increasing order:
1.82, 1.92, 1.96, 1.98, 2.00, 2.05, and 2.10.

The hypothesis is that 1.82 is the outlier. The mean is 1.98.

The number closest to 1.82 is 1.92.
Therefore, the gap is $1.82 - 1.92 = 0.1$

The range is the maximum value minus the minimum value.

The range is $2.10 - 1.82 = 0.28$
$Q = 0.1/0.28 = 0.36$

The $Q_{critical}$ for eight values are: $Q_{90\%} = 0.468$; $Q_{95\%} = 0.526$, $Q_{99\%} = 0.634$

$$0.36 < 0.468$$

Even at the 90 percent confidence level, the data point cannot be rejected. We are really stuck with 1.82!

[1] $G(\alpha, n)$ values for Grubbs' test were obtained from ChemWiki maintained by The University of California Davis. http://chemwiki.ucdavis.edu/Reference/Reference_Tables/Analytic_References/Appendix_07%3A_Critical_Values_for_Grubb%E2%80%99s_Test, accessed May 21, 2015.

CHAPTER 13
Optical Spectroscopy

SECTION A

The Basics of Concentration Determination

BACKGROUND AND THEORY

This chapter provides a very brief introduction to the basics of ultraviolet-visible (UV/Vis) spectroscopy. The theory of how and why molecules absorb light in this range is extremely interesting and much deeper than presented in this short introduction. Detail can be found in general and organic chemistry texts and additional facets in physical chemistry texts.

With many biological molecules, concentration measurements are accomplished using ultraviolet or visible absorbance spectroscopy. The predominance of UV/Vis spectroscopy in the life sciences is because it is fairly inexpensive and that it can provide accurate information for biomolecules as well as many small organics. UV/Vis benchtop units can be bought for a few thousand dollars and visible-only units for less than $1000. Of course, with advanced features like auto-sampling and advanced computer control, much more can be spent.

In the ultraviolet and visible ranges of the electromagnetic spectrum, compounds that contain π or nonbonding electrons absorb light. The incident light emitted by the appropriate lamp in the spectrophotometer causes these more loosely bound electrons to enter an excited state – a highest occupied molecular orbital (HOMO)-to- lowest unoccupied molecular orbital (LUMO) transition. The detector measures the light that passes through or is transmitted by the sample. For UV/Visible spectroscopy, the wavelengths of light used are in the ultraviolet and visible regions of the electromagnetic spectrum.

If light is absorbed by the sample, the light hitting the detector is reduced.

$$A = \log I_0/I$$

A is the absorbance, I_0 is the incident light intensity, and I is the intensity of light exiting the sample.

Light that is not absorbed is transmitted.

$$T = I/I_0$$

Transmittance, T, is often expressed as a percent and is related to absorbance.

$$A = -\log T = -\log I/I_0$$

(Refer to logarithms in Chapter 4 to do the math.) Since A and T are ratios, they are unitless quantities.

Practical UV/Visible spectroscopy is generally employed between 220 nm and 700 nm of light. In this range, molecules with extended π systems, found in conjugated bonds, absorb—often strongly. Like proteins, biomolecules contain the aromatic amino acids tryptophan, phenylalanine, and tyrosine; nucleic acids contain the aromatic bases adenine, guanine, cytosine, thymine, and uracil. These molecules impart strong absorption of light in the ultraviolet range. As the degree of conjugation increases, the gap between the HOMO and LUMO decreases, and lower-energy (longer wavelength) light is necessary to cause an excited state. The energy of transition moves into the visible range, and compounds absorbing in this range are colored. Biomolecules like the bright orange β-carotene exhibit strong visible absorbance.

What makes this absorption so critical is that the quantity of light absorbed is generally directly proportional to the number of molecules absorbing the light. Additionally, the specific wavelengths of light absorbed are characteristic of the compound absorbing the light.

SECTION B

The Beer-Lambert Law: Theory, Applications, and Limitations

The Beer-Lambert law relates the absorbance of a solution to its concentration.

$$A = \varepsilon bc$$

Where A is the absorbance of the solution, ε and b are constants and c is the concentration (generally expressed in M).

The absorbance, A, is defined as above:

$$A = \log I_0/I$$

The constant, ε, molar absorptivity, is also known as the extinction coefficient and is constant for a given molecule at a given wavelength. The most common unit for ε is $Lmol^{-1}cm^{-1}$. Values for ε range from as low as hundreds of thousands. The stronger the absorbance, the higher the molar absorptivity. The final constant, b, is the path length and is a parameter of the instrument. For convenience, the spectrophotometer is usually designed so that the path length is 1 cm.

This proportional relationship between absorbance and concentration allows a researcher to carefully prepare a series of known concentrations of a molecule of interest (often using serial dilution; see Chapter 7) and measure the absorbance of each solution. Since the Beer-Lambert law has a linear form:

$$A = \varepsilon bc + 0$$
$$y = mx + b$$

A plot of A versus c will produce a line with a slope equal to εb. The y-intercept, b in $y = mx + b$, should pass through the origin (0,0) and give a y-intercept of 0. A solution with zero concentration should produce zero absorbance. However, small variations from the origin are commonly seen. In fact, the deviation of the intercept from zero is a measure of the quality of the data. Additionally, if the plot is fit with a linear trend line, the R^2 value will also address data quality. With careful work, R^2 values of 0.999 can be realized. Once the slope is established, the constant can be applied to determine the concentration of any unknown solution of the same molecule if the absorbance of that solution is measured.

Limitations of Beer's Law

As powerful as the Beer-Lambert law (commonly and more amusingly called Beer's law) is, it does have limitations. These limitations can be chemical (as a result of the molecular nature of the analyte) or instrumental (as a result of limitations of the spectrophotometer). Although the limitations can be discussed in very technical terms, a practical list is given here.

Light scattering caused by particulates within the sample or light emission by the sample in the form of fluorescence or phosphorescence will cause a nonlinear relationship between absorbance and concentration. At very low concentrations, the threshold of detection may not provide accurate results. This threshold will vary, depending on the sensitivity of the instrument and the molar absorptivity of the analyte. Strongly absorbing compounds can be detected at much lower concentrations than those with small ε values. At high concentrations (sometimes at concentrations >0.01 M), electrostatic interactions between molecules in a solution, changes in refractive index, or shifts in chemical equilibrium affect absorbance and cause deviations in linearity. Stray light within the instrument can also cause deviations from linearity and are an artifact of the instrument.

In the Laboratory

Example 13-1:

Riboflavin, also known as vitamin B-12, produces an intense orange yellow solution when dissolved in ethanol. A 0.1 mM solution of standard riboflavin is prepared in ethanol. Five serial dilutions are prepared by pipetting 1.00 mL of the stock to a final volume of 2.5 mL. The absorbance values, measured at 266.5 nm, were 3.251, 1.370, 0.527, 0.223, 0.091, and 0.034. An unknown solution of riboflavin shows an absorbance of 0.872. Determine the molarity of the unknown solution of riboflavin.

The concentrations of each sample must be calculated. This can be accomplished easily given the constant dilution factor, df = 1/2.5 or 0.4 (if you do not remember serial dilutions or dilution factors, see Chapter 7). The data are summarized in Table 13-1.

TABLE 13-1: Absorbance Values and Calculated Concentrations for Riboflavin in Ethanol, Measured at 266.5nm. Concentrations Must Be Converted from mM to M (See Chapter 5).

Conc. (M)	A
1.00E-04	3.251
4.00E-05	1.370
1.60E-05	0.527
6.40E-06	0.223
2.56E-06	0.091
1.02E-06	0.034

To determine the concentration of the unknown, the molar absorptivity for riboflavin must be determined. This requires a plot of absorbance versus concentration as shown in Figure 13-1.

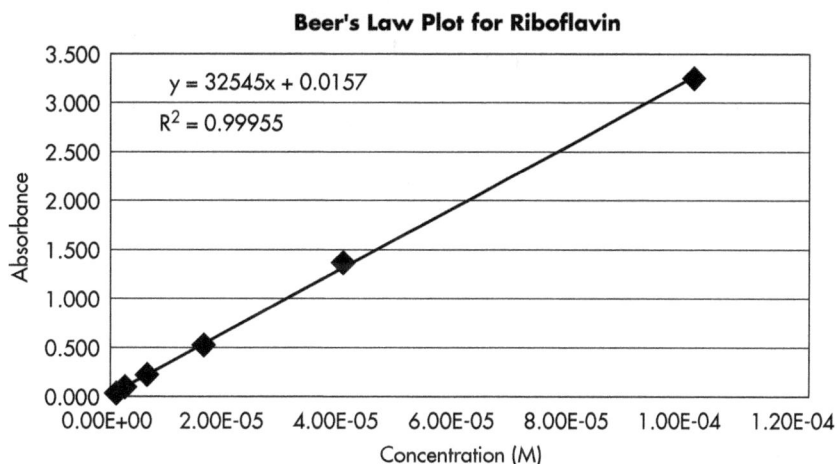

FIGURE 13-1 Beer's law plot for riboflavin. The data in Table 13-1 were plotted and fitted with a linear trend line. The equation of the line is $y = 32{,}545x + 0.0157$. The slope (m = εb) is 32,545 cm/M; the coefficient of determination is 0.99955.

Using Beer's law:

$$A = \varepsilon bc$$

And substituting the slope, 32,545 cm/M, for εb and the absorbance 0.872 for A:

$$0.872 = 32{,}545 \text{ M}^{-1} \times c$$
$$c = 2.68 \times 10^{-5} \text{ M}$$

The deviation of the y intercept from 0 is small and the R^2 is high, indicating little deviation from linearity.

Reality Check: The absorbance of the unknown solution is 0.872, between the second and third sample, concentrations of 4.0×10^{-5} M and 1.6×10^{-5} M. The concentration of 2.68×10^{-5} M falls between those two values. The concentration is reasonable.

GLOSSARY

Accuracy—Accuracy refers to how close the experimental measurement is to the true measurement for the sample. Accuracy is most simply expressed as percent error: (experimental value—theoretical value)/theoretical value $\times$ 100.

Acid—Acids come in three types: Arrhenius acids, Brønsted-Lowry acids, and Lewis acids. For solutions, the Brønsted-Lowry definition is the most appropriate. A Brønsted-Lowry acid is a substance that can donate a proton (H^+) to a Brønsted-Lowry base.

Activity (a)—Activity is a measure of the effective concentration in solution. Activity is most appropriately applied in real (nonideal) solutions. Given the difficulty in measuring activities, concentrations are commonly substituted for the more correct activity measurement.

Activity coefficient (γ)—The activity coefficient is a factor that measures a deviation of a chemical substance from ideality. The activity coefficient multiplied by the molar concentration produces the activity of a component in solution.

Amphoteric—Amphoteric refers to the ability of a compound to act as both an acid and a base. Water is the most common amphoteric substance, as it can act as an acid producing hydroxide ion (OH^-) or as a base producing hydronium ion (H_3O^+).

Avogadro's number—Avogadro's number is $6.022140 \ldots \times 10^{23}$. It is the number of particles, most often molecules or atoms, which are contained in one mole of a substance.

Base—Bases come in three types, Arrhenius bases, Brønsted-Lowry bases, and Lewis bases. For solutions, the Brønsted-Lowry definition is the most appropriate. A Brønsted-Lowry base is a substance that accepts a proton (H^+) to a Brønsted-Lowry base.

Beer-Lambert law—The Beer-Lambert law, commonly known as Beer's law, is used in optical spectroscopy to mathematically relate the optical absorbance of a compound to its molar concentration.

$$A = EbC$$

where A is the absorbance of light, E is a constant called molar absorptivity or extinction coefficient and C is the molar concentration of the substance. Beer's law fails in highly concentrated or highly dilute solutions.

Biological unit (IU or U)—A unit (IU or U) is a quantity of a biologically active substance like an enzyme, hormone, or vitamin required to produce a specific response. Units are specific to a particular reaction or set of reactions or particular effect. An enzyme unit (U) is the amount of a particular enzyme that catalyzes the conversion of 1 micromole of substrate per minute. The IU originates in pharmacology: one IU is equal to one milligram.

Buffer—A buffer is a solution that resists changes in pH when diluted or when subject to additions to acids or bases. Buffers are composed of a weak acid and its conjugate base (salt) or a weak base and its conjugate acid (salt).

Buret—A buret is a piece of chemical glassware used for careful measurement of liquid volumes, usually in titrations. Burets contain volume lines (gradations) that allow measurement of the volume of liquid (generally water) delivered. Burets may be accurate to the hundredths place.

Common logarithm—A logarithm with the base 10. Generally, a logarithm is a common logarithm unless another base is specified.

Conjugate acid—In Brønsted-Lowry acid-base theory, a conjugate acid is the product of the reaction of the base. It is the member of the acid-base pair that has accepted a proton (H^+) from the acid.

Conjugate base—In Brønsted-Lowry acid-base theory, a conjugate base is the product of the reaction of the acid. It is the member of the acid-base pair that remains after the acid has donated a proton (H^+).

CRC Handbook—*The CRC Handbook of Chemistry and Physics* is a nearly exhaustive reference book of all registered chemicals and includes vast tables of constants and information on nomenclature, symbols, and units. It is available in both print and online versions. Its strength lies in its completeness; its weakness is that chemical knowledge is required to search in the print version.

Dalton (Da)—The dalton is one unit of mass on an atomic or molecular scale. It is defined as 1/12 the mass of carbon-12 nucleus and is the same as one atomic mass unit (amu). One dalton is numerically equivalent to 1 g/mol. Daltons are commonly used as a unit for biological molecules.

Density—The density of a substance is the ratio of its mass to its volume. Most commonly, the units employed are g/mL.

Dibasic—Dibasic refers to an acid that donates two hydronium ions in an acid-base reaction. It can also refer to a salt that has two univalent (+1) metal ions. H_2SO_4 is a dibasic acid. Na_2HPO_4 is a dibasic salt.

Dilution equation—The dilution equation relates the concentration of a stock solution to a diluted solution. Its most general form is $C_1V_1 = C_2V_2$ with the requirement that the concentrations C_1 and C_2 have identical units. The same restriction holds in that V_1 and V_2 must also have identical units. A more specific form of the equation is $M_1V_1 = M_2V_2$ where M refers to molar concentration.

Dilution factor—The dilution factor is the ratio of the initial concentration of a solution to its final concentration.

Dimensional analysis—In chemical sciences, dimensional analysis is a method used to convert one measurement to another measurement: 60 seconds equated to 1 minute is an example of dimensional analysis.

Equilibrium—Equilibrium in a chemical reaction is the point where the concentrations of reactants and products do not change. The reaction still proceeds; however, the rates of the forward and reverse reactions are equal.

Extinction coefficient—See **molar absorptivity**.

Formula mass—The formula mass, generally expressed using the units g/mol, is the sum of the atomic masses of each atom in a formula unit of a compound; for water (H_2O), the formula mass of 18.0 g/mol is equal to the mass of two hydrogen atoms (2 × 1.0 g/mol) and one oxygen atom (16.0 g/mol).

Formula weight—See **formula mass**.

Graduated cylinder—A graduated cylinder is a piece of chemical glassware used for routine measurement of liquid volumes. Graduated cylinders contain volume lines (gradations) that allow measurement of the volume of liquid (generally water) contained therein. Graduated cylinders may be accurate to the hundredths place.

Henderson-Hasselbalch equation—The Henderson-Hasselbalch equation is an approximation (simplification) of the equilibrium expression for the dissociation of a weak acid or a weak base. It is used to conveniently predict buffer composition (acid and base ratio) or pH for a buffer solution. It has a number of forms, but one of the most common for a weak acid (HA)/salt (A^-) solution is $pH = pKa + \log [A^-]/[HA]$.

Heterogeneous—Heterogeneous refers to more than one phase. In a heterogeneous solution, distinct layers or separation of species can be observed.

Homogeneous—Homogeneous refers to a single phase. In a homogeneous solution, no distinct layers or separation can be observed.

Hydrate—A hydrate is a chemical compound where one or more water molecules is included in the molecular structure. Hydrates are noted with the number of water molecules as a suffix to the formula. The molecular formula $CuSO_4 \cdot 5H_2O$ indicates five water molecules as part of the chemical formula of the molecule. In determining the formula mass of a hydrate, the mass of the water molecules must be included.

Hydronium (hydronium ion)—Hydronium ion is the conjugate acid of water. It is positively charged with the formula H_3O^+. It is the strongest acid that can exist in aqueous solution. It is often abbreviated as a proton (H^+); however, H_3O^+ is the correct form in water.

Hydroxide (hydroxide ion)—Hydroxide ion is the conjugate base of water. It is negatively charged with the formula OH^-. It is the strongest base that can exist in aqueous solution.

Immiscible—Immiscible refers to liquids that are not soluble, forming layers or a heterogeneous solution.

Kw—Kw is the ion product constant for water. It is the result of the product of the molar hydronium ion concentration multiplied by the molar hydroxide ion concentration. At 25°C, it is 1.00×10^{-14}. $Kw = [H_3O^+][OH^-]$.

Logarithm (log)—A logarithm represents the power (exponent) to which a number (the base) must be raised to produce a given number. Any base can be used; however, the base 10 (see common logarithm) and base e (see natural logarithm) are the most commonly used bases.

Measurement—A measurement is the assignment of a numerical value to an object with defining units: 0.01 is a number; 0.01 mL is a measurement.

Merck Index—The *Merck Index* is a reference book providing names, structures, and basic physical information about chemicals, drugs, and biological molecules. It is available in print and online versions. Its strength lies in its simple alphabetical listing by compound name.

Micropipettor—See **pipettor**.

Miscible—Miscible refers to liquids that are mutually soluble, forming a homogeneous solution.

Molality (*m*)—Molality is a unit for expressing the concentration of a solution. It is the moles of solute per kg of solvent (mol/kg). Molality is not temperature dependent.

Molar absorptivity—Molar absorptivity, also called a molar extinction coefficient, is a constant that quantifies how strongly a substance absorbs light. It is measured for a single compound at a defined wavelength. Common units are $Lmol^{-1}cm^{-1}$.

Molar mass—See **formula mass**.

Molarity (M)—Molarity is a unit for expressing the concentration of a solution. It is the moles of solute per liter of solution (mol/L). Since the volume of solution is used, molarity is a temperature-dependent measurement.

Mole (mol)—A mole is the amount of pure substance that contains the same number of chemical units (atoms or molecules) as there are in exactly 12 grams of carbon-12. In practical terms it is defined as Avogadro's number (6.022×10^{23}) of atoms or molecules. Mole is often confused with molecular mass, which is the mass of one mole of atoms or molecules, and has the units g/mol.

Molecular weight—See **formula mass**.

Monobasic—Monobasic refers to an acid that donates only one hydronium ion in an acid-base reaction. It can also refer to a salt that has only one univalent (+1) metal ion. HCl is a monobasic acid. NaH_2PO_4 is a monobasic salt.

MSDS—MSDS is an abbreviation for Material Safety Data Sheets. These documents contain information on health and safety data for chemical and some biological compounds. MSDS sheets are published by the manufacturer or seller of chemicals.

Natural logarithm (ln)—A logarithm with the base *e*. The base *e* is 2.71828…

Normality (N)—Normality is a unit for expressing the concentration of a solution. It is based on equivalent mass and uses the units of equivalents/L (eq/L). It is commonly used in acid-base chemistry where one equivalent means one mole of acid (H^+) or one mole of base (OH^-). Since the volume of solution is used, normality is a temperature-dependent measurement.

Percent (%)—Percent is a unitless quantity composed of one measurement in one hundred like measurements: 1 percent is one part in 100 parts (pph).

pH—pH (*pondus hydrogenii*) is a scale used to conveniently express the hydronium ion (acid) concentration in solution. The pH is the negative base 10 logarithm of the molar concentration of hydronium ion in solution $-pH = -\log[H_3O^+]$. As acid concentration increases, the pH decreases.

Pipette—A pipette, more appropriately termed a volumetric pipette, is a laboratory tool used for delivering carefully measured (up to four significant figures) volumes of liquid (generally water). Pipettes may be fixed volume, appropriate for measuring only one volume; or variable, containing gradations.

Pipettor—A pipettor, also called a micropipettor, uses air displacement generated by a piston to provide suction to draw a carefully calibrated volume of solution from a vessel. Micropipettors can achieve accuracy up to four significant figures. Pipettors may be fixed volume, appropriate for measuring only one volume; or variable, allowing for the selection of an appropriate volume.

pKa—pKa is an expression of the strength of an acid. It is the negative base 10 logarithm of the acid dissociation constant. The log scale is used to more conveniently express the equilibrium values: pKa = $-$ log Ka. As the acid strength increases, the pKa decreases.

pKb—pKb is an expression of the strength of a base. It is the negative base 10 logarithm of the base dissociation constant. The log scale is used to more conveniently express the equilibrium values: pKb = $-$ log Kb. As the base strength increases, the pKb decreases.

pKw—pKw is the negative base 10 logarithm of the Kw (ion product constant for water). Its value at 25°C is 14.

ppb—ppb is the abbreviation for parts per billion. It is a unitless quantity composed of one measurement in one billion like measurements: 1 ppb can represent $1\ g/10^9\ g$ or $1\ mL/10^9\ mL$.

ppm—ppm is the abbreviation for parts per million. It is a unitless quantity composed of one measurement in one million like measurements: 1 ppm can represent $1\ g/10^6\ g$ or $1\ mL/10^6\ ml$.

Precision—Precision refers to the variability in or closeness of measurements in a set or series of identical measurements. The simplest expression of precision is range, which is the highest value less the lowest value. A common expression of precision is standard deviation (s or σ).

Salt—Salt is the common name for the ionic compound composed of the conjugate base of an acid together with its cation or the conjugate acid of a base together with its anion. Salts are electrically neutral as they possess equal cationic and anionic charges. Salts are produced as the result of a neutralization reaction.

Serial dilution—Serial dilution is the stepwise dilution of a stock solution. Often, the dilution at each step is constant. Each subsequent diluent solution decreases in concentration; from a 1 M, two serial dilutions of 1 mL to 10 mL would provide an intermediate concentration of 0.1 M, and, diluted from the intermediate, a final concentration of 0.01 M.

SI units—SI stands for *Système Internationale d'Unités*, the standard international system of measurements. The SI base units are m, kg, s, K, A, mol, and cd.

Significant figures—Significant figures are the number of digits that carry meaning in a measurement. Significant figure rules provide a basis to ascertain what digits are known with certainty.

Solute—A solute is the component of a solution that is present in the smaller amount in a solution. A solute is dissolved in a solvent. Solutes can be solid, liquid, or gaseous.

Solution—A solution is a homogeneous mixture of two or more substances. Solutions may be solid, liquid, or gaseous.

Solvent—A solvent is the component of a solution that is present in the greatest amount. Solvents can be liquid or gaseous.

STP—STP refers to standard temperature and pressure. Standard temperature is 0°C and pressure, 1 atm.

Strong acid—A strong acid is an acid that dissociates (donates H^+) completely. Therefore, at equilibrium, there is no appreciable amount of the intact acid present. Instead, the conjugate base concentration is equal to the initial acid concentration.

Strong base—A strong base is a base that is completely dissociated (ionized in solution). Thus, at equilibrium, there is no appreciable amount of the base present. Rather, the conjugate acid concentration is equal to the initial base concentration.

Temperature—Temperature is a measure of the heat content in a system. It is not heat. The most common units for temperature are Celsius (°C), Fahrenheit (°F), or Kelvin (K).

To contain (TC)—*To contain* is one category of two types of glassware. In TC instruments, the glassware must be drained or completely transferred to achieve the calibrated volume. TC pipettes should be blown out or otherwise have the remaining liquid removed for accurate volumes.

To deliver (TD)—*To deliver* is one category of two types of glassware. In TD glassware, the small volume of liquid that remains in the glassware after draining has been taken into account in the calibration of the instrument. Therefore, TD pipettes should not be blown out or otherwise have the remaining liquid removed. Such action would result in a high volume delivered.

Tribasic—Tribasic refers to an acid that donates three hydronium ions in an acid-base reaction. It can also refer to a salt that has three univalent (+1) metal ions. H_3PO_4 is a tribasic acid. Na_3PO_4 is a tribasic salt.

Unit—A unit is any standard used for a comparison of measurements. A number together with its unit constitutes a measurement.

Unit cancellation—Unit cancellation is a method for conducting dimensional analysis where any unit divided by the identical unit results in a value of one: mL/mL = 1 is an example of unit cancellation.

Unit flipping—In unit flipping, any measurement can be expressed in more than one manner, as long as the numerical value remains associated with the unit: 1 g/2 mL is equivalent to 2 mL/1 g is an example of unit flipping.

Volumetric flask—A volumetric flask is a vessel calibrated to measure a fixed volume of liquid (generally water) with a high degree of accuracy (relative uncertainty of up to 400 ppm).

Weak acid—A weak acid is an acid that dissociates (donates H^+) incompletely. Therefore, at equilibrium, there are appreciable amounts of the intact acid and its conjugate base present.

Weak base—A weak base is a base that does not fully ionize in solution: its protonation is incomplete. Hence, at equilibrium, there are appreciable amounts of the base and its conjugate acid present.

www.ingramcontent.com/pod-product-compliance
Lightning Source LLC
Chambersburg PA
CBHW081536220326
41598CB00036B/6450